SpringerBriefs in Computer Science

Series Editors

Stan Zdonik, Brown University, Providence, RI, USA
Shashi Shekhar, University of Minnesota, Minneapolis, MN, USA
Xindong Wu, University of Vermont, Burlington, VT, USA
Lakhmi C. Jain, University of South Australia, Adelaide, SA, Australia
David Padua, University of Illinois Urbana Champaign, Urbana, IL, USA
Xuemin Sherman Shen, University of Waterloo, Waterloo, ON, Canada
Borko Furht, Florida Atlantic University, Boca Raton, FL, USA
V. S. Subrahmanian, Department of Computer Science, University of Maryland,
College Park, MD, USA
Martial Hebert, Carnegie Mellon University, Pittsburgh, PA, USA
Katsushi Ikeuchi, Meguro-ku, University of Tokyo, Tokyo, Japan
Bruno Siciliano, Dipartimento di Ingegneria Elettrica e delle Tecnologie
dell'Informazione, Università di Napoli Federico II, Napoli, Italy
Sushil Jajodia, George Mason University, Fairfax, VA, USA
Newton Lee, Institute for Education, Research, and Scholarships, Los Angeles, CA,
USA

More information about this series at http://www.springer.com/series/10028

Stefan Gruner • Apurva Kumar • Tom Maibaum
Markus Roggenbach

On the Construction
of Engineering Handbooks

with an Illustration from
the Railway Safety Domain

 Springer

Stefan Gruner
Department of Computer Science
University of Pretoria
Pretoria, South Africa

Tom Maibaum
Computing and Software
McMaster University
Wadhurst, East Sussex, UK

Apurva Kumar
Kitchener, ON, Canada

Markus Roggenbach
Computer Science
Swansea University
Swansea, UK

ISSN 2191-5768 ISSN 2191-5776 (electronic)
SpringerBriefs in Computer Science
ISBN 978-3-030-44647-5 ISBN 978-3-030-44648-2 (eBook)
https://doi.org/10.1007/978-3-030-44648-2

This Springer imprint is published by the registered company Springer Nature Switzerland AG
The registered company address is: Gewerbestrasse 11, 6330 Cham, Switzerland

Foreword

Among the earliest engineering handbooks are those of the Greek authors Ktesibios, Philon and Heron on the making of catapults (c. 350 BC). They had formulæ that enabled makers to scale the catapult to the ammunition with a view to maximising the range and impact. Their texts address practical technologies and respond to pressing needs of the day, not least the subject of the even earlier handbook of Aineias Tacticus *How to survive under siege*. Their books are a remarkable testimony to the sophistication of Greek technology, upon which the Romans built so effectively. Writing down of a body of knowledge that can be learned, transmitted and used in the world's work is not new.[1]

Today, the world's work is everywhere technical and professional. It is supported by curricula, qualifications, accreditation, charters, and —when particularly new, in development or transformation— by concepts of best practice, communities and regulation. The invasion of software into everything has created many software engineering domains that are self-sustaining with their own bodies of knowledge and practices, e.g., in science, engineering, business, entertainment and transport. The pace and scale of new developments suggest that the formulation and writing down of a body of knowledge and practice that can be used in specialised domains of software engineering is invaluable and necessary.

This little book tackles questions about the nature of engineering knowledge and practice through the concept of the practitioners' handbook. The conception of a handbook is demanding as its purpose is to seek and record what knowledge and know how really matters and to make it usable. The authors of this book combine ideas from many quarters and reflect on how to identify and organise 'settled' knowledge into a handbook, in particular regarding our capabilities for making software that is safe, secure, resilient,

---

[1] The catapult is an extraordinary case study when thinking about the nature of engineering knowledge; see T.E. Rihll, *The Catapult*, Westholme Publishing, 2007; and T.E. Rihll, *Technology and Society in the Ancient Greek and Roman Worlds*, The American Historical Association and the Society for the History of Technology, 2013.

etc. The roots, inspiration and focus of the book are in Formal Methods for software engineering. The distilled handbook production approach however applies to engineering in general. Their exemplar is of a domain of software engineering that is coming of age, the design and implementation of software based signalling in railways. The authors' perspective is novel, and their discussion is revealing and stimulating. They offer us an intriguing new way of thinking about contemporary technological knowledge.

Swansea, February 2020   *John V. Tucker*

Preface

Established engineering disciplines (and, in a similar way, the discipline of medicine) have desktop handbooks which are partly descriptive and partly normative: they give the practitioners of those disciplines a systematic overview of their disciplines' knowledge, which comprises both topic knowledge (about the objects which are in the scope of the discipline), as well as method knowledge (about how to proceed in order to solve problems which often and typically arise within those scopes). For newly 'emerging' domains or disciplines, however, for which no Handbook (HB) with normative authority has yet been defined, the question arises of how to do this systematically and in a non arbitrary manner; this is the focus of this book.

Its meta methodological tasks entail: clarification of what actually a 'HB' is, the systematic identification of what ought to be considered as 'settled' knowledge (extracted from historic repositories) for inclusion into such a HB, and the 'assembly' of such identified knowledge into a form which is 'fit' for the purpose and conforms to the formal characteristics of HBs as a 'literary genre'.

This book is the first to reflect upon the question of how to construct a desktop HB. 'Settled knowledge' is defined and identified as the key ingredient for HB production. It is demonstrated how concept analysis can be used for identifying settled knowledge by utilizing the assembled data for classification; a presentation scheme for HB articles is developed and demonstrated to be suitable.

Modern society increasingly utilizes computer systems, presuming them to be dependable, i.e., safe and secure. Railway control is a typical example. Computer Science needs to address the challenges of (1) designing dependable systems and (2) providing evidence for safety and security properties of such, often complex, systems. Formal Methods are one important means to address both of these questions. However, actual Formal Methods HBs are scarce or nonexistent. A HB would encourage and enable practitioners to use Formal Methods as an applicable, every day tool for software development.

   This book is rooted in the philosophy and methodology of engineering. It provides a clear definition of settled knowledge and concise presentation of methodologies for HB development, exemplified in the railway domain. These cover the question of how to identify settled knowledge and also of how to transform such identified knowledge into a set of informative handbook articles. Finally, the limitations of these methods are discussed.

   With the recently emerging 'discipline' of 'Formal Methods in the railway domain' as our motivating and illustrative example, this book shows, in principle, how a HB can be reasonably constructed.

Pretoria, Kitchener, Wadhurst, Swansea   *Stefan Gruner*
February 2020   *Apurva Kumar*
                                                          Tom Maibaum
                                                    Markus Roggenbach

Acknowledgements

We thank *John V. Tucker* for writing a foreword to our book, and also for giving us excellent feedback. Many thanks to *Jan Peleska*, who provided valuable feedback on an early draft of this book. Thanks also to *Jonathan Bowen* and *Silvano dal Zilio* for their valuable advice about some of the taxonomic classifications recapitulated in chapter 2. Further relevant literature hints were received from *Hubert Garavel* and *Jaco Ackermann*. Special thanks to *Erwin R. Catesbeiana* for keeping us on track. Last but not least, many thanks to *Ralf Gerstner* from Springer-Verlag for the professional curation of our book in this edition.

Contents

Part III : SYNTHESIS

5   Example HB Entry of a Formal Method for the Railway Domain — Step 6 57

6   Conclusions and Prospects for Future Work 77

Part I
Background

Chapter 1
Introduction and Motivation

Practitioners of well established applied sciences, which include the old discipline of medicine, as well as mature engineering disciplines, possess desktop *handbooks* (HBs), which contain the trustworthy and widely accepted 'recipes' for practice in those disciplines. HBs give the practitioners of those disciplines a systematic overview of their disciplines' contents. HBs comprise both *topical knowledge* about the entities which are in the scope of the discipline, as well as *methodological knowledge* about how to proceed to solve the problems which typically arise within those scopes. HBs in the applied sciences (and, in particular, the engineering disciplines) thus include both *descriptive* and *normative* elements in some analogy to the 'paradigmatic textbooks' which the historian and philosopher of science, Thomas Kuhn, has characterised as constitutive and school forming for the classical scientific disciplines [58].

The purpose of HBs is to offer codified methods. These constrain the creativity of the practitioner so that what is built has some likelihood of working (being 'fit for purpose') and being safe. In other words, codified methods provide a clear route to guaranteed and repeatable success.

The word 'handbook' (or 'manual') has its roots in the classical Greek word ενχειριδιον (encheiridion). Such manuals already existed in times of antiquity. Some of the oldest manuals were of a rather 'philosophical' bent. For example, the *Enchiridion of Epiktet* provided practical moral advice on the theoretical basis of Stoic ethics. Others were technical manuals for the artisan crafts, devoted to more 'practical' matters, such as to the making of catapults. Common to these manuals is that they contained instructions about *how to* do something properly, an important feature of a 'handbook' in our understanding.

Whereas the handbooks of historically old scholarly disciplines (e.g., medicine) have been 'growing organically' (metaphorically speaking) since time immemorial, the situation today is fundamentally different with its 'mushrooming' of ever newer academic and practical disciplines and sub-disciplines at a historically rapid pace. In these cases, we are no longer in

S. Gruner et al., *On the Construction of Engineering Handbooks*, SpringerBriefs
in Computer Science, https://doi.org/10.1007/978-3-030-44648-2_1

the position to hope and wait for the 'organic growth' of the handbooks for those new disciplines 'as time goes by'; instead, we must construct them purposefully at a pace which matches the rapid 'pace of emergence' of those new disciplines and subdisciplines. One example of such an emerging discipline is that of *Formal Methods in the railway domain*; it has existed for approximately 20 to 30 years. However, a normative and authoritative handbook on this 'young' topic is not yet available. For a purpose like this (as well as many similar ones), we need a rational HB construction method (actually a meta method because HBs must contain method knowledge) that is effective and efficient, as well as being justifiable methodologically and in terms of the philosophy of science.

Such a meta method of HB construction, which is the topic of this book, must address at least the following questions, which are together, in their totality, both *historic* and *systematic*, as well as both *descriptive* and *normative*:

- Clarification of the fundamental question, *what* actually *is* a 'HB', in contrast to what is actually *not* a HB. In other words, we need to establish *criteria* that distinguish HBs from non HBs.
- Then we need to be able to determine *which* 'elements' of 'knowledge', out of a historically vast repository of documented knowledge, may or shall be included into a new HB to be developed and established. Choices will have to be made, because a HB must necessarily be smaller than a discipline's entire historic repository.

  - Because a HB is intended to be used 'daily' by the practitioners of a discipline, the HB's contents must be considered by those practitioners as *trustworthy* and *reliable*.
  - Trustworthiness, however, can only be ascribed to knowledge and methods which have already to some extent passed the 'test of time' again and again, such that they can henceforth be regarded as *settled knowledge*, which will likely continue to remain trustworthy for years to come, notwithstanding the usual research progress.
  - Hence the meta method must deal with the questions of *where to find* and *how to identify* 'settled knowledge' in a discipline's historic repository. This, again, requires *criteria* which must appear as 'reasonable' from a philosophy of science point of view.

- Last but not least: After the 'settled knowledge' has been identified and chosen for inclusion into a new HB (previous point), and after it has been clarified what a HB must internally 'look like' to make it acceptable as a HB (first point), the meta method must also provide guidelines about how to *transform* the textual and structural *presentation* of the settled knowledge into a *form* that makes it HB-*compatible*.

In this book, we address these elements, present steps that *can* be carried out and *show by example how this can be done*. In other words: we devise a meta method and demonstrate (by example) that it is, in general, effective and feasible. The chosen domain, from which our example is 'extracted', is the already mentioned discipline of 'Application of Formal Methods in Railway Engineering'.

The key contribution of this book is the (example illustrated) construction method itself, *not* the HB which would result from a highly detailed and very comprehensive application of our method.

1.1 What constitutes a HB?

According to Jackson [45] an *engineering handbook*, which we consider to be a HB, is *not* merely a compendium of fundamental principles. Instead, it contains a corpus of *rules* and *procedures* by means of which those fundamental principles can be most easily and efficiently *applied* to the design tasks which typically occur in the chosen area of engineering. Thus, the 'outline' of a 'typical' design is almost always already given, as it is predetermined by the already existing products, in conjunction with their corresponding human needs and requirements [45]. Indeed, according to Maibaum [69], this is what engineering bodies of knowledge *ought to be* about.

Peculiar 'historical, semantic shifts', in which the meanings of key terms are gradually changing over time, are typical for all advanced sciences during the course of their history [27]. Though software engineering is still a comparatively young discipline, it already has experienced some semantic shifts of its key terms [33]. This is a problem that makers, as well as readers, of HBs must always keep in mind.

For the sake of clarification, we provide the following two *counter examples* of what we do *not* regard as HBs in the above mentioned sense:[1]

- The well known SWEBOK, in its 3rd edition from 2014,[2], presents in its more than 300 pages merely *factual* knowledge (like an encyclopedia), but strongly lacks applicable *methodical* knowledge in the form of typical problem solution 'recipes'. By analogy we might say that the SWEBOK is like a comprehensive treatise about all sorts of flour, eggs, milk, sugar, and many further ingredients, however, *without* telling its readers anything about *how to* bake a cake.
- Oliveira [72] provides a comprehensive survey about which European universities have taught what kinds of Formal Methods in their Computer Science curricula. Although such a survey is without doubt a noble undertaking, which can serve many other useful purposes, it does not help any practitioner to 'look up' the 'most recommended' solution S for some

---

[1] Pointers to these two counter examples were provided to us by some of our academic colleagues during the discussion phase of this book's manuscript prior to its publication.

[2] https://www.computer.org/web/swebok

given Formal Methods problem P in the context of a chosen application domain (e.g., the railway domain).

Engineering disciplines [84], as well as other science based and application oriented disciplines such as medicine, are characterised by *high levels of standardisation* and the subsequent availability of readily applicable handbook knowledge. In those engineering or applied science disciplines, a handbook is comprised of:

- factual knowledge: explanations of concepts used;
- procedural knowledge: how we can go about doing things;
- problem classification: grouping problems with similar characteristics;
- design methods: recipes for designing solutions for problems.

These are elements that characterise a professional domain. Such HBs can be found on the book shelf or on the desk of every serious practitioner. *Positive examples* of HBs that contain all these elements (to different degrees) are:

- H. Llewelyn, H.A. Ang, K. Lewis, A. Al-Abdullah, *Oxford Handbook of Clinical Diagnosis*. Oxford University Press, Third Edition, 2014.
- R.C. Dorf (ed.), *The Engineering Handbook*, CRC Press, 2005.
- I.N. Bronshtein, K.A. Semendyayev, G. Musiol, H. Mühlig, *Handbook of Mathematics*. Springer, 2015.

From a philosophy of science point of view, we can state that the contents of such handbooks are often presented in the 'nomopragmatic' logical form [14][3], on the basis of which even more detailed 'technological rules' can be formulated. In those application oriented contexts,

- *'nomological'* statements describe lawful scientific facts on which all proper engineering must be grounded;
- the above mentioned *nomopragmatic statements* describe science based technical *possibilities* or options;
- whereas, the *technological rules* outline the adequate actual *implementations* of those possibilities [14].

Therefore, *only 'settled'* knowledge is trustworthy enough for inclusion into any professional engineering HB. Seen from the point of view of the philosophy of engineering, the notion of 'settledness' (and, hence, trustworthiness) of engineering knowledge is closely related to Vincenti's notion of *'normal design'* [84], in analogy to Kuhn's well known notion of *'normal* science' [58]. In the context of engineering, 'normality' entails *"the improvement of the accepted tradition, or its application under new or more stringent conditions"* [84]. An implied (though not explicitly stated) view of 'normal' engineering design is that engineers normally design *devices* (as opposed to *systems*). A

---

[3] Gr.: *nomos* = law, rule, custom; *pragmatic* relates to acting or doing. An example of a nomopragmatic statement provided by Bunge is: *"If a magnetized body is heated above its Curie point, then it is demagnetized"* [14](p. 149).

device, in this sense, is an entity the design principles of which are already well known, well defined, well structured, and subject to the principles and conditions of 'normal' design. A 'system', on the contrary, which is a subject of so called '*radical* design', is an entity which *lacks* several of those characteristics which *would* make its 'normal' design possible. This notion of 'radical design' in engineering corresponds quite well to Kuhn's notion of 'science *in crisis*' [58] in the field of the theoretical (non applied) sciences. Examples of 'devices' given by Vincenti [84] are conventional airplanes, electric generators, turret lathes, and the like, whereas examples of 'systems' are airline companies, electric power plants, or automobile factories as a whole.[4] It would thus appear that 'systems' become 'devices' when their design attains the status of being 'normal', i.e., when the problem of 'original creativity' required for their design becomes merely a matter of systematic choice, based on well defined analysis, in the context of standard definitions and criteria developed and accepted by the 'community' of engineers, in an analytical industrial manner. Further details about Vincenti's relevance for Software Engineering and Formal Methods (along the lines of what we have briefly sketched above) can be found in [43] [67] [68].

With similar ideas in mind, a theorist of Informatics, Dirk Siefkes, had spoken about so called '*small* systems' which have much in common with what Vincenti had called 'devices'. In [79], Siefkes called a system "*small, if it is in every aspect adequate, not exaggerated, and not too little*". In other words, what Vincenti's 'devices' and Siefke's 'small systems' have in common (as seen from a philosophy of technology point of view) is that they both refer to something that is reasonably 'familiar', hence manageable, and thus '*makes sense' for us* in a pragmatic, hermeneutical manner. Both notions are obviously related to the ontological notion of '*Zeug*' (i.e., 'equipment') by the phenomenologist philosopher Martin Heidegger in his seminal *Sein und Zeit* (Being and Time) [39]. Thus, all things of 'Zeug' type populate our medium scale '*meso' world.* 'Zeug' is thus neither too large nor too small for us, since we grow up with 'Zeug' in a quasi natural manner prior to any deeper philosophical analysis of it [39]. Hence it must be such a kind of 'Zeug' which is represented by the contents of a HB in the practical engineering realm. Valuable and 'HB-ifiable' methods in the realm of engineering are therefore the '*micro* methods' on the basis of '*micro* theories' [43] that are (again) closely related to the tasks of developing 'devices' in Vincenti's sense of the term. Accordingly, it is a highly important objective of (software) engineering research to construct and compile useful 'catalogues' of such 'micro methods' in support of the daily work of the (software) engineering practitioner. *Such a 'catalogue' is a HB* according to our new understanding of this term. In other

---

[4] For the sake of illustration, Whereas we have factories that produce cars almost entirely robotically, with little human intervention on its production lines, we do not (yet) have any meta factory to robotically produce and emit car factories on the basis of the design plans of the cars which these factories are meant to produce. Thus, whereas the car is already a device, its factory is not (yet) a device, given the current 'state of the art'.

words, unlike the 'grand theories' to which the classical theoretical sciences aspire, HB oriented 'micro theories' describe only small and well 'manageable' parts of some domain of interest, such that they become immediately 'useful' in any development phases of 'normal' engineering projects [43].

1.2 Settled Knowledge

Poser [74], Vincenti [84], as well as Arageorgis and Baltas [3], described *engineering* as a multi level activity that is both analytic and constructive. Implied are various categories of knowledge, all of which are relevant in almost every engineering project [84]. Those categories include *explicit* forms of knowledge, such as 'fundamental design concepts', 'criteria', and 'theoretical tools' (e.g., Formal Methods), as well as *tacit* knowledge (which is much harder to identify), such as general 'guidelines', 'rules of thumb', traditional 'tricks of the trade', and the like [84]. In this context, 'settled' knowledge is the kind of knowledge which *can* become 'officially' codified in a HB. Its structure would thus be consistent with Vincenti's general epistemological categories [84], as well as with domain specific application purposes. Such 'officially' codified knowledge must be stable and coherent throughout a HB's topic range, over a reasonably long period of time, and ought to appear in similar form for all similar problems to be solved in that domain. *Forms* of engineering knowledge could thus include mathematical formulae, semi formal descriptions of applicable methods, or even pictures and diagrams, as long as they are clearly problem solution oriented. Settled knowledge in those (and further) forms typically stems from the 'knowledge generating activities' which can be found comprehensively described in [84].

For our purposes in this book, we adopt a working definition of settled knowledge that enables us to proceed with our investigation. Such a working definition is not intended to be definitive, in the sense of being 'settled', but is sufficiently precise and scientifically plausible to form a basis for further work. Like any definition, it should enable precise analysis of the elements of ideas that follow from it. It is subject to revision as we learn more about the domain of study. Justified by the paragraph above, and supported by the considerations below, we give the following

Working Definition of 'Settled Knowledge':   Some piece of knowledge is *settled knowledge* in a given domain if and only if it is an element of normal design for that domain (see section 1.1).

In addition to Vincenti's considerations concerning the epistemology of engineering, we can find further methodological support in the classical philosophy of science, especially applied science, by Mario Bunge [14]. An important notion therein is the notion of *operational theories*, which are meta scientific theories about scientific methods, rather than scientific theories about natural objects [14]. For HB building, specifically in the field of applied computer science (including software engineering), which is a highly abstract discipline that hardly possesses any 'natural' objects, operational theories are of prominent importance. For the sake of any future HB in our chosen domain, it is thus especially important to identify and describe the relevant operational theories. For the purpose of HB building, we must thus survey large amounts of literature sources, spread out over a reasonably long period of publication time, in a search for the most applicable (and most often reported) operational theories. These operational theories are, in the case of our extended example, the Formal Methods of computer science (as they are applied in the railway domain).

1.3 Formal Methods in Design and Validation of Railway Control Systems

Formal Methods in software science and software engineering have existed at least as long as the term 'software engineering' (from the NATO Science Conference, Garmisch, 1968) itself. In many engineering based application areas, especially the domain of railway engineering (which also includes inner city tram lines, urban monorail systems, and the like), Formal Methods have reached a level of maturity that already calls for the compilation of a HB. The domain's various methods and techniques include: algebraic specification, process-algebraic modeling and verification, Petri nets, fuzzy logic, and several more. The B method, for example, has already been used successfully for the correct development of the most relevant software components of the Metro underground railway system of the city of Paris (France). This successful example is indicative of the maturity and 'HB worthiness' of the Formal Method applied in it.

According to Fantechi's recent overview [26], *"the verification of complex railway signalling systems is still a main challenge"* and entails *"an important percentage of the cost in the development of these systems"* [26](p. 168). The term 'Formal Methods' is, thus, broadly understood in the railway domain as encompassing *"all notations having a precise mathematical semantics, together with their associated analysis and development methods, that allow to describe and reason about the behaviour and functionality of a system in a formal manner, with the aim to produce an implementation of the system that is provably free from defects"* [26](p. 168).

Within the broader railway domain, in particular the subdomain of *"signalling has been traditionally considered as one of the most fruitful areas of intervention for formal methods"* [26](p. 169), with an impressive array of industrial success cases [26](pp. 169-170).[5] These success cases often include the model checking of interlocking systems [26](pp. 170-171). Examples include [6] [48] [49]; see also [37] for a comparative study. In contrast, theorem proving techniques are, by and large, not yet extensively applied in the railway domain [26](p. 172). Model based design and development in the railway domain is already well known and widely accepted, so that Petri nets, state charts, and finite state machines are reported by Fantechi as especially popular techniques [26](p. 173).

Those classical software modelling techniques, however, are increasingly confronted with latest developments in railway engineering, such as (for example) the shift in automated train protection (ATP) from static fixed block to dynamic moving block track segmentation, as well as the introduction of on train wireless communication systems, by means of which moving trains can automatically 'see' (and 'talk to') each other [26](p. 175). At this point, abstract software modelling is still 'limping one step behind' hardware engineering and must 'catch up' with the hardware's latest features and possibilities. Further software modelling and interfacing problems in this domain are also due to the quickly increasing integration of, hitherto, rather isolated technical subsystems, such as (for example) interlocking and ATP [26](p. 176). Moreover, in addition to the classical *safety* properties found in these topics, *liveness* properties are becoming increasingly important [26](pp. 179-180), too.[6] *"Hence, the key is to adopt a multi level modelling approach to address the complexity of a railway system"*, Fantechi concluded [26](p. 180).

1.4 Structure of the Remainder of this Book

Chapter 2 on 'related work' concludes this 'background' part I. Chapters 3 and 4 constitute the 'analysis' part II of this book. Chapters 5 and 6, finally, bring everything together in the 'synthesis' part III. The contents of this book's parts II and III is briefly outlined in the following paragraphs.

As our handbook composition method comprises of *six steps* in total, we provide an overview of these steps; we identify 'ideal roles', describe the different types of expertise required to carry out the different steps and characterise the results of each of these steps. Later we discuss each step in more detail.

---

[5] By contrast, a train collision with eleven casualties between Holzkirchen and Rosenheim in Bavaria (Germany), 9-Feb-2016, had been caused by a human operator who had manually *switched off* the automatic train control (ATC) system; that accident was thus not due to engineering faults nor technical defects.

[6] Note that, according to [2], every possible specification can be expressed as the conjunction of a safety and a liveness property.

Thereafter, we provide an *extended example*, namely of how to apply our general method to the railway domain. This example is covered by two chapters:

- Chapter 4 is devoted to illustrating the *first five steps* of our method. These are of a mainly analytical nature, namely on identifying and classifying 'settled' knowledge. We conclude Chapter 4 with a critical evaluation of steps 1 to 5 in the context of the Railway Example.
- Chapter 5 illustrates the *sixth, final 'transformational step'*, which we recommend for the 'HB-ification' (i.e., HB compatible textual and structural representation) of the selected knowledge. To this end, we provide a sketch of a handbook entry on the topic: *Verification through Model Checking of Ladder Logic Programs for Safety.* Such an entry would belong to the 'Traffic Monitoring and Control' area of the railway domain, which we identify in Chapter 4 as one of the areas where our extended example reveals the existence of settled knowledge.

Note, however, that chapters 4 and 5 are only illustrative; we do *not* claim to have already applied our HB construction method 'exhaustively', to the extent that we could immediately start the construction of an actual HB. The construction of a handbook for Formal Methods in the railway domain remains future work, for which the collaboration of many experts would be needed. Nonetheless, we conjecture that our methodological considerations in this book will make any future HB composition project more likely to succeed; it serves as a 'meta recipe' about how to write 'recipe books'.

Chapter 6 concludes our book by highlighting its main contributions and by outlining our view of the *future work* that remains to be done.

Chapter 2
Related Work

With respect to the purpose of this book, we consider as 'related work':

- publications that discuss the organisation of engineering knowledge,
- publications in which some systematic methods for (or approaches to) HB construction (especially *Formal Methods* HB construction), are shown, as well as
- publications in which some classifications of Formal Methods into various categories are presented.

Accordingly, some relevant publications in these three fields are briefly recapitulated in the subsequent three sections.

2.1 Organisation of Engineering Knowledge

"*Engineering refers to the practice of organising the design and construction of any artifice which transforms the physical world around us to meet some recognised need*" [76].

The nature, purpose and practices of engineering lead directly to the need for engineering HBs. Rogers' definition, quoted above, entails that engineering is different from science and mathematics. The differing objectives and methods of the three disciplines induce important differences in practice. Moreover, they induce significant differences in the education of practitioners. Finally, conventional engineers are taught the principles of the 'devices' they use in designing artefacts as well as the systematic design principles to be used in building instances of such devices. In [76], Rogers explained in great detail how engineering is different from science. He argues this view based on what he called the 'teleological distinction', which concerns the differing aims of science and technology:

> "*In its effort to explain phenomena, a scientific investigation can wander at will as unforeseen results suggest new paths to follow. Moreover, such investigations never*

© The Author(s), under exclusive license to Springer Nature Switzerland AG 2020
S. Gruner et al., *On the Construction of Engineering Handbooks*, SpringerBriefs in Computer Science, https://doi.org/10.1007/978-3-030-44648-2_2

end because they always throw up further questions. The essence of technological investigation is that they are directed towards serving the process of designing and manufacturing or constructing particular things whose purpose has been clearly defined. We may wish to design a bridge that uses less material, build a dam that is safer, improve the efficiency of a power station, travel faster on the railways, and so on. A technological investigation is in this sense more prescribed than a scientific investigation. It is also more limited, in that it may end when it has led to an adequate solution of a technical problem. The investigation may be restarted if there is renewed interest in the product, either because of changing social or economic circumstances or because favourable developments in a neighbouring technology make a new advance possible. On the other hand, it may come to a complete stop because the product has been entirely superseded by something else that will meet humanity's changing needs rather better".

Moreover, we quote from the same author:

"We have seen that in one sense science progresses by virtue of discovering circumstances in which a hitherto acceptable hypothesis is falsified, and that scientists actively pursue this situation. Because of the catastrophic consequences of engineering failures, whether it be human catastrophe for the customer or economic catastrophe for the firm, engineers and technologists must try to avoid falsification of their theories. Their aim is to undertake sufficient research on a laboratory scale to extend the theories so that they cover the foreseeable changes in the variables called for by a new conception. The scientist seeks revolutionary change, for which he may receive a Nobel Prize. The engineer too seeks revolutionary conceptions by which he can make his name, but he knows his ideas will not be taken up unless they can be realised using a level of technology not far removed from the existing level" [76](p. 55).

Vincenti defined engineering activities slightly differently w.r.t the *design, production* and *operation* of artefacts. Of these, especially design and operation are highly pertinent to software engineering, while it is often argued that production plays a very small role, if any, in the practice of software engineering.[1] In discussing the focus of engineers' activities, Vincenti also mentioned the notion of '*normal design*',[2] which comprises *"the improvement of the accepted tradition or its application under new or more stringent conditions"*. The *"engineer engaged in such design knows at the outset how the device in question works, what are its customary features, and that, if properly designed along such lines, it has good likelihood of accomplishing the desired task"* [84](p. 7).

In the context of *Software Engineering*, Michael Jackson discussed such a concept of 'normal design', too [44], though he initially did not use this phrase himself:[3]

1. *"In this context, design innovation is exceptional. Only once in a thousand car designs does the designer depart from the accepted structures by an*

---

[1] Once a software system 'stands', it can be (re)produced effortlessly as many times as needed simply by copying its code text.

[2] The term originated in [21].

[3] Jackson adopted Vincenti's terminology only in later publications.

innovation like front-wheel drive or a transversely positioned engine. True, when a radical innovation proves successful it becomes a standard design choice for later engineers. But these design choices are then made at a higher level than that of the working engineer: the product characteristics they imply soon become well understood, and their selection becomes as much a matter of marketing as of design technology. Unsuccessful innovations, like the rotary internal combustion engine, never become established as possible design choices".[4]

2. *"An engineering handbook is not a compendium of fundamental principles; but it does contain a corpus of rules and procedures by which it has been found that those principles can be most easily and effectively applied to the particular design tasks established in the field. The outline design is already given, determined by the established needs and products".*

3. *"The methods of value are micro-methods, closely tailored to the tasks of developing particular well-understood parts of particular well-understood products".*

Another important aspect of engineering design is the organising principle of hierarchical design [44]: *"Design, apart from being normal or radical, is also multilevel and hierarchical. Interesting levels of design exist, depending on the nature of the immediate design task, the identity of some component of the device, or the engineering discipline required".* An implied (though not explicitly stated) view of engineering design is that engineers normally design *'devices'* (as opposed to 'systems').

Hence our pursuit of HBs in the domain of Software Engineering is entirely consistent with the general practice in Engineering.

2.2 Other Approaches to HB Construction

In spite of the many already existing handbooks in and for a plethora of fields, and in spite of the great demand for such HBs on nearly every topic, little has been published so far about the methods along the lines of which such HBs ought to be systematically composed; in other words, *'meta HBs'* on the topic of HB writing are still lacking in most fields and disciplines.

According to Taguchi (et al.), who briefly sketched their ideas on 'building a body of knowledge on model checking for software development' in [81], a *body of knowledge* (BOK) *"is a collection of substantial concepts and skills that represent knowledge of a certain area in engineering/scientific discipline, and ensures its common understanding. A BOK may include technical terms*

---

[4] Note, however, that the Wankel engine mentioned above by Jackson had been successfully used in many Mazda cars, even including a race victory at Le Mans in 1991, and that the company recently issued a statement according to which this type of motor will be used again as a range extender in a forthcoming generation of electric battery vehicles; see https://www2.mazda.com/en/publicity/release/2018/201810/181002a.html

and theoretical concepts as well as recommended practices" [81](pp. 784-785). The authors of this paper first identified several 'knowledge areas' $(KA_1, \ldots)$ that were then *"further classified into subareas up to four tiers"*. Alas, in [81] we cannot find much more than a descriptive listing of those knowledge areas, i.e., no rationale for the detection and identification of 'settled' knowledge, no stepwise transformation of chosen knowledge into applicable, problem solution oriented technical rules, and the like.

Also, Bowen and Reeves have taken a rather traditional vantage point from which they characterised a BOK, somewhat vaguely, as *"an ontology for a particular professional domain"* [13](p. 308). In the most salient parts of their paper, Bowen and Reeves described the various aspects of the relationship between a BOK and its supporting 'community of practice'. As a result, they also had to start their considerations with the observation that a proper and useful HB for the domain of Formal Methods does not yet exist [13](p. 309), in spite of Formal Methods having *"now reached a level of maturity when an associated Body of Knowledge would be a worthwhile part of the general effort to ensure that Formal Methods find their rightful place in the software engineering profession"* [13](p. 321). Most interesting in the approach of Bowen and Reeves is their attempt at meta specifying the appropriate structure of a Formal Methods BOK by means of Formal Methods themselves (in their case the formal language Z), for the sake of clarity. Alas, the Z specification of a BOK's envisaged structure in [13](pp. 317-319) makes it clear that Bowen and Reeves pursue a rather 'encyclopedic' approach, according to which a BOK is not substantially more than an internally hyperlinked dictionary similar to the 'Wikipedia'[5] pages on the Internet. Neither have Bowen and Reeves emphasised in [13] the importance of applicable, problem solution oriented 'cook book recipes', though it was vaguely mentioned that knowledge *"would include the appropriateness of various combinations in different situations"* [13](p. 321); nor have they indicated any rational criteria by means of which the actual contents of an envisaged BOK ought to be selected and extracted from an already existing literature database or knowledge repository.

Most recently, a systematic method for gathering and classifying information about relevant themes, trends, and subdomains of the railway domain from literature resources was published in [57], without, however, our specific focus on applicable Formal Methods, and also not with any explicit plan to write a HB on this topic. Nonetheless their bibliometric, statistical classification and ranking approach might well be useful for the identification of 'settled knowledge', which is an important 'work package' for the writers of a HB in any field.

---

[5] https://www.wikipedia.org/

2.3 Other Classifications of Formal Methods

In spite of the long history of Formal Methods research and development [12] [30], surprisingly little has been published so far about systematic (or taxonomic) classifications (or categorisations) of the Formal Methods which are already known. The same is true, more specifically, for Formal Methods in the railway domain. The few already existing taxonomies or classifications of Formal Methods tend to be rather coarse, e.g., 'logic-based', 'algebraic', and the like, and are typically not application domain specific. A recent ACM computer science topic classification scheme (2012),[6], for example, simply divides 'Formal Methods' into four broad branches, namely, 'Model checking', 'Software verification', 'Automated static analysis', and 'Dynamic analysis', which are neither domain exhausting nor pairwise disjoint.

The oldest Formal Methods classification attempt, of which we are aware, dates to the year 1990 and can be found in [87]. It includes attributes such as 'tool supported' versus 'not tool supported', and several more. Though it is *"not the intent of this book to give a complete taxonomy of all possible characteristics of a method, nor to classify exhaustively all methods according to these characteristics"* [87], several distinguishing criteria were nevertheless provided, including: 'model-oriented' versus 'property-oriented' Formal Methods, where a system's behaviour is specified 'directly' versus 'indirectly'.[7]. Furthermore, 'visual' Formal Methods, whose languages contain graphical elements, were distinguished from purely textual ones, as well as 'executable' Formal Methods from non executable ones [87]. Several example formalisms to match those categories were also described.

In [20], published in 1996, several Formal Methods were grouped w.r.t. their typical *intentions* in relation to applications, e.g., for *"specifying the behaviour of sequential systems"*, versus *"specifying the behaviour of concurrent systems"* [20]. 'Hybrid combinations' were also mentioned, *"one for handling rich state spaces and one for handling complexity due to concurrency"* [20]. Moreover, 'model checking' and 'theorem proving' were mentioned as the two most typical verification approaches; 'classical' mathematical engineering methods (e.g., equation solving on the basis of numerical models) were not listed as prominent. Several industrial examples to match those categories of Formal Methods were also provided, with the understanding that *"no one formal method is likely to be suitable for describing and analysing every aspect of a complex system"* [20].

Another rather early Formal Methods classification attempt was published by Axel van Lamsweerde [63] in the year 2000. He focused particularly on formal *specification* (rather than verification) techniques, and distinguished between 'history based', 'state based', 'transition based', 'functional', and 'operational' formal specifications. The functional specifications were further

---

[6] http://www.acm.org/about/class/2012

[7] In other words: 'explicitly' versus 'implicitly'.

subdivided into 'algebraic' and 'higher order' specifications [63]; several examples were provided and discussed.

In 2008, the difference between methods of 'static analysis' and 'dynamic analysis' was briefly mentioned in [40], which served the purpose of a broad non technical overview paper for software practitioners. Therein, 'abstract interpretation' was mentioned as an important meta method with applicability in several concrete Formal Methods. Rather vaguely, 'weak' Formal Methods (without proof systems) were distinguished from 'strong' Formal Methods (with proof systems), as well as 'heavyweight' versus 'lightweight' Formal Methods (which distinction relates to their subjectively perceived difficulty of practical applicability) [40].

A 'taxonomic approach' to classifying Formal Methods specifically for the application domain of web services was attempted in [18] in 2010. Alas, this project was prematurely aborted and did not produce more than a few brief and vague suggestions. A small classification graph was shown in [18](figure 5: p. 360); unfortunately, that graph does not explain anything beyond what was already known from other literature. In its conclusion, the paper merely stated "that a taxonomy is needed for guiding the combination and usage of formal methods" [18](p. 360). As in our own approach, however, formal concept analysis (FCA) was proposed in [18] as a step towards achieving such classifications. Like ours, the classification attempt sketched in [18] was also domain-specific, in contrast to most of the other Formal Methods classification attempts, which are unrelated to specific application domains.

In the same year, the need for a classification 'ontology' for the domain of verification and validation (V&V) was identified in [55]: "Given the huge number of V&V methods and tools, a formal description of the V&V domain is mandatory in order to ease the choice of the best technology when a developer needs to verify and/or validate parts of the system he/she is building". The approach published in [55] "represents a knowledge sharing initiative for the V&V domain and provides a formal representation of the key data regarding this domain". Thus, the theoretical intention and the practical purpose of the V&V project reported in [55] are quite similar to the intention and purpose of our work in the railway domain, though the authors of [55] "have chosen the Web Ontology Language (OWL)" for their purpose instead of formal concept analysis (FCA).[8] In one of their illustrative examples, Petri nets were classified further into 'algebraic', 'decision free', 'timed' and 'consistent' nets [55]. The "Related work" of [55](sect. 4) states that "to the best of our knowledge, there is no ontology reported anywhere for the V&V domain", which coincides with our own impression concerning the scarcity of classification literature in the Formal Methods (FM) domain, in general, as well as for the railway domain (FM-Rail), in particular.

---

[8] Here it should be noted that OWL and FCA have different purposes and semantics: OWL is a knowledge representation language that supports (logical) analysis, while FCA is a concept modelling language.

More recently, in 2014, 'specification oriented' Formal Methods were distinguished from 'analysis oriented' ones in [11]. This distinction is, by and large, the same as the previously mentioned distinction between 'weak' and 'strong' Formal Methods, as in [40]. In [11], furthermore, 'algebraic' Formal Methods were distinguished from 'model oriented' ones, see the above mentioned paper [87] for comparison, where the 'algebraic' methods were also called 'property oriented' in [11]. (However, [87] was neither mentioned nor cited by [11].)

Specifically for the *railway* domain, we can find a recent classification of often applied (mostly state based and/or algebraic) Formal Methods in [70], which must be additionally augmented with the various methods (based on logic programming) described in [66], and Petri nets [83].

A recent list of Formal Methods which are frequently used in the *transport* domain (which includes the railway domain) can be found in [5]; such summaries can be useful sources for the identification of 'settled knowledge' on which every reliable HB must be based. The relevance of Formal Methods for the (general) transportation domain is also highlighted in [31](figure 10), though it was not indicated which Formal Methods specifically are the most important ones for this domain [31].

The use of Formal Concept Analysis (FCA) for many different classification purposes in various application domains, which is also our approach, is already well known and much described in the scientific literature. Several examples, for the sake of illustration, can be found in [7] [42] [71] [75] [86].

2.4 Comparison with our Approach

In most of the related work mentioned above, existing formal specification or verification languages or methods were grouped and categorised based on their own intrinsic features and properties. We could thus call those related work approaches '*intrinsic*' classification attempts. Our classification approach, by contrast, can be characterised as '*extrinsic*' because it groups Formal Methods according to their *occurrence* (or usage) in particular application projects as they were reported by the various sources from the railway domain. An 'extrinsic' property of a Formal Method in our classification approach is thus a domain-specific application relation, *not* an inherent feature of a Formal Method itself. This conceptual distinction between 'intrinsic' and 'extrinsic' schemes of classification must be taken into account methodologically, in order not to misunderstand the purpose and the results of our ongoing project, which we had already previously sketched methodologically in [36]. The purpose of our classification scheme, as already explained in [36], is to identify and distinguish whether some particular Formal Methods are arguably 'settled', i.e., already 'mature enough' based on rationally justifiable criteria, especially w.r.t. their trustworthy applicability in the railway domain.

Another extrinsic classification attempt for Formal Methods can be found in [30](ch. 2: pp. 43-65), dating from 2013; there, however, we can once more merely find a list of various application domains without more specific information about which particular Formal Methods are most suitable in those domains.

Our (meta) method of HB composition strongly relies on classifying Formal Methods into 'settled' or 'not settled' with the help of the mathematical notion of 'stability' in *formal concept lattices*. In this context, last but not least, one more 'related work' ought to be mentioned in which methodological recommendations concerning stability analysis had been provided [22] in 2010. However, whereas formal concepts with *low* stability were regarded as 'indicative' and noteworthy in [22] (supported by an argument according to which low stability might point to intrinsic conceptual design flaws of the underlying formal concept lattice itself), our work, as indicated already in [36], does *not* regard concepts of low stability as particularly noteworthy. We are interested only in concepts with *high* stability. Thereby we assume (in contrast to the considerations of [22]) that our underlying formal concept lattice itself i,s by and large error free, i.e., 'error free' w.r.t. our empirical observations in the data base which lead to the design of our lattice scheme. 'Deeper' discussions concerning the degrees of 'interestingness' of formal concepts situated in such lattices can be found most recently in [62].

Part II
Analysis

Chapter 3
A General Method for Composing an Engineering HB

As is also the case in other engineering disciplines, software engineering knowledge can be organised in the categories of engineering knowledge identified by Vincenti [84]:

1. Fundamental Design Concepts;
2. Criteria and Specifications;
3. Theoretical Tools;
4. Quantitative Data;
5. Practical Considerations;
6. Design Instrumentalities.

Fundamental design concepts include the *operational principle* of their device (i.e., how the engineered entity at the heart of the solution to a problem works). According to Polanyi this means *"knowing for a device how its characteristic parts ... fulfill their special functions in combining to an overall operation which achieves the purpose"* [73]. A second principle typically taken for granted is the *normal configuration* of the device, i.e., the commonly accepted arrangement of the constituent parts of the device. These two principles (and possibly others) provide a framework within which normal design takes place. *Criteria and specifications* allow the engineer to translate general, qualitative goals into concrete technical terms. This translation requires the device to follow a given operational principle and to be in a normal configuration. The development of such criteria may be difficult. However, the development and acceptance of such criteria is an inherent part of the development of engineering disciplines.

Engineers require *theoretical tools* to underpin their work. This includes intellectual concepts for thinking about design, as well as mathematical methods and theories for making design calculations. Both conceptual tools and mathematical tools may be devised specifically for use by the engineer and be of no particular value to a scientist or mathematician. Indeed, as it was stated by Jackson, *"the most useful context for the precision and reliabil-*

ity that formality can offer is in sharply focused micro methods supporting specialised small scale tasks of analysis and detailed design" [45].

Engineers also use *quantitative data*, as well as tabulations of functions in mathematical models. A typical example in the field of software engineering can be found in Knuth's compendium of sorting and searching [56].

There are also *practical considerations* in engineering. These are not usually subject to systematisation in the sense of the categories above, but reflect more pragmatic concerns. For example, a designer will have to make various trade offs that are the result of general knowledge about the envisaged device, its use, its context, its cost, etc.

Design instrumentalities include *"the procedures, ways of thinking, and judgmental skills by which it is done"* [84]. This is related to what the well known 'capability maturity model' (CMM) intends where it refers to well defined and repeatable processes in software engineering.

According to Vincenti, as noted above, the daily activities of engineers consist of normal design as *"the improvement of the accepted tradition or its application under new or more stringent conditions"*. This is the combination of discipline and a limited amount of creativity encapsulated in engineering 'cookbooks': *"The engineer engaged in such design knows at the outset how the device in question works, what are its customary features, and that, if properly designed along such lines, it has a good likelihood of accomplishing the desired task"*.

In our view, a HB organises such engineering knowledge in a form that enables an engineer to recognise the problem of concern, to identify potential solutions of this problem, and to systematically implement an acceptable solution. In the following parts of this chapter we present a method along the lines of which such a HB can be produced.

To carry out our HB production method, different types of expertise are required in the various production steps. This expertise can best be described by way of *'ideal roles'*. Any such role can be played by an individual or a group, i.e., we do *not* require a bijective (1:1) mapping between these functional roles and the human experts who play these roles and serve in their functions. These roles are briefly characterised as follows (in *no* particular order of importance or priority).

Domain Expert:
  knows and understands the application domain for which a HB is to be written and *can* also act as 'editor in chief' for an entire HB production project.

Formal Concept Analysis Expert (FCA Expert):
  knows how to carry out formal concept analysis (FCA) on a given data set.

Scientific Reader:
  also understands the prospective application domain and is, moreover, able to identify, to retrieve and to collect relevant data, as well as relevant scientific or technical literature on the topic of interest.

Author:

has specialized technical knowledge in a specific part of the domain and will be mainly responsible for writing a HB chapter (or section) about this topic.

Editor:

coordinates (like a 'project manager') the activities of all other role players in an envisioned HB's production process.

The steps of our method and the expertise required to carry them out are outlined briefly in the subsequent paragraphs; further details and explanations follow throughout this book.

Step 1:
Choice of Sources for Settled Knowledge

In order to identify settled knowledge, a suitable corpus of literature about the domain under discussion, from both academic and technical sources, must be identified and studied. The expertise required in this step must be sufficiently deep and wide, so as to ensure sufficiently comprehensive coverage of the domain. The above mentioned *Domain Expert* will be strongly involved in this step.

Result:

Corpus of relevant literature.

Step 2:
Choice of Domain Specific Attributes and Data Collection for Formal Concept Analysis (FCA)

This is the first concrete step towards building a domain model. The domain specific attributes constitute the terminological basis over which we will build a conceptual model of our domain of interest. These attributes are identified in a 'dialectical' process. On the one hand, the *Scientific Reader* identifies and highlights terms and notions that appear frequently (de facto) in the corpus produced in Step 1. On the other hand, the *Domain Expert* provides terms and notions that are *normatively expected* to appear in the envisaged HB's 'universe of discourse'. This dialectic process leads to questions like: Which terms and notions shall remain? Another problem is the development of *univocal* terminologies, as the corpus of Step 1 might contain different words for the same entity. In such cases the *Domain Expert* ought to be able to identify them. Similarly, there might also be closely related terms and notations that are, strictly speaking, not equivocations, but which the Domain Expert would nonetheless wish to consider as 'equivalent' descriptors of one and the same entity at a slightly higher level of generalisation.

Result:
Objects and attributes discussed in the corpus.

Step 3:
Application of FCA to the Data Collected

This step 'organises' the domain on the basis of the concepts, objects, and attributes identified in the corpus. To this end, the above mentioned *FCA Expert* uses tools to automatically compute a so called *concept lattice* and a graphical visualisation of it. Nodes and partial order relations in this lattice are results of a theory based computation (i.e., not a deliberate act of creativity by the FCA Expert using a tool). Lattice nodes can represent objects, attributes and concepts. The lattice's partial order illustrates the relationships between the domain's attributes and objects, objects and concepts, as well as the generalisation relationship between concepts and their sub or super concepts.

Result:
Concept lattice that captures all relevant findings from the corpus.

Step 4:
Choice of Stability Threshold

In this step, the *FCA Expert* uses built in statistical methods of the FCA tool to 'prune' the concept lattice and to 'clear' it of 'noisy' information of lesser importance. The theory of FCA provides an application domain *independent* solution for this problem. The act of pruning focuses the expert's attention onto lattice elements with 'high stability'; these seem to be central to the domain of interest about which the envisaged HB shall be produced.

Result:
Pruned and 'stable' concept lattice from which 'noisy' information of lesser importance has been removed.

Step 5:
Classification of Settled Knowledge

In this step, the *Domain Expert* evaluates if the previous (mostly automatic steps) provide a satisfactory result that seems to *'make sense'*. The subjective view of the Domain Expert and the algorithmic corpus analysis results are thus supplemented and adjusted. In this step, two main problems must be considered:

- The corpus chosen in step 1 itself might have been somehow insufficient. Such a flaw might have biased all the subsequent analysis results. For example, there might be some concepts or even entire subdomains which are not properly represented at all. In such a case, the domain expert must expand the corpus (by including previously omitted data or literature sources) and then recompute the entire concept lattice.
- The domain expert might have had a personally biased opinion about the domain that cannot be supported by any given corpus. In such a case, the expert ought to 'learn the lesson' from observation and adjust any inadequate or outdated personal opinions.

Result:

Consolidated representation of the 'settled knowledge' in the domain of interest, including the relationships between the 'elements' of this settled knowledge, which might already be regarded as a first version of the 'Table of Contents' (albeit in no particular sequential order) of the envisaged HB (possibly after several iterations).

Step 6:
Presentation of Settled Knowledge

The table of contents (albeit in no particular sequential order) and the structure of the HB arise from the results of the classification exercise sketched above. However, the various HB entries ought to follow a standardised presentation scheme which should adequately reflect the nature of 'normal engineering methods' [84]. Vincenti did not himself present such a standard, though he aptly described the 'components' of such standardised knowledge. The chosen scheme must subsequently be 'filled in' by the above mentioned Authors who have the requisite expertise on the specific topics to write their corresponding HB entries.

Result:

Handbook (HB).

After the foregoing sketchy overview of the six steps, the finer details of each of them are discussed in the subsequent sections.

3.1 Step 1: Choice of Sources for Settled Knowledge

Choosing appropriate sources of settled knowledge is crucial for the success of constructing reliable HBs. In order to do this, one must first investigate where the settled knowledge for a chosen domain of interest may 'live'. Several factors are important for making this decision. Not only should such a knowledge source be actively and successfully used in the domain; it should also already have been used for a reasonably long period of time.[1] Duration of use allows the separation of settled knowledge from what is merely the 'flavour of the month' or with what the community of experts is still only 'tentatively experimenting'.

One might start by *interviewing senior experts* currently practicing in the domain of interest; for comparison, see [4].[2] The experts' long term professional knowledge might rightfully be considered as 'settled' since the techniques and principles they apply are used to solve practical problems encountered in their domains on a daily basis, with good results. However, each practitioner might still have a unique 'personal' approach to the typically given problems. While large portions of the relevant knowledge can be assumed to 'overlap' amongst the individual members of a community of experts, there may nonetheless be significant variations depending on the type of problems an expert is exposed to. However, visiting several experts in order to gather their combined knowledge may be problematic as well, as this approach considerably increases the time needed for information retrieval and for checking that the information thus retrieved is consistent (free of mutual contradictions). Moreover: even a mutually consistent collection of several expert opinions provides us only with a *current* 'snap shot' of domain knowledge, such that this information retrieval procedure might have to be repeated more than once, over an extended time period, in order to obtain an overview of that knowledge that is sufficiently 'long lasting' for inclusion into the envisaged HB. These activities are closely related to what Bjørner had called 'domain engineering' in one of his books [10](pp. 7-8).

Documented knowledge (whether officially published or informally 'in circulation') might thus be a more feasible starting point for the necessary information retrieval. Examples of such domain specific documents include:

- Industry standards and guidelines laid out by governing bodies;
- Papers and articles published as a result of scientific research in the domain;
- Requirements documents produced by domain experts and specialists, in and for 'typical', as well as sufficiently large and significant 'real world' projects.

But which of these would provide the 'best' resource for 'settled' knowledge? Knowledge expressed in official standards or guidelines is likely settled. However, these official guidelines are often presented at rather high levels of generality in order to leave enough room for situation specific implementations

---

[1] Scientists are most strongly interested in what is *new*, while engineers are more conservatively interested predominantly in what 'works well'.

[2] According to the questionnaire based survey of [4], most applications of Formal methods in industrial railway projects focus on interlocking systems; Formal Methods are mainly used for formal specification and formal verification in the early requirements and design phases; the B family is the most popular tool suite; the most relevant functionalities are formal verification and support for formal modelling; the most relevant quality features are related to the maturity and usability of the tools [4](p. 770).

and problem solutions. Hence we cannot automatically expect such official guidelines to be very detailed, such that additional sources of more detailed knowledge must still be found. Moreover we must keep in mind that even official standardisation documents can become outdated and are, thus, regularly updated and replaced by newer versions. Hence we would also need to examine different (earlier versus later) versions of the same standardisation documents in order to find out which knowledge in them is 'stable' across a longer duration of time that spans several 'versions' of such documents.

Scientific conferences are mainly focused the newest pieces of information (at the 'cutting edge' of research) in a domain as opposed to well established approaches to common problems in that domain. This raises the question of whether or not scientific conferences are the right place to find what constitutes settled knowledge. However, some really fundamental concepts and notions also have long term 'stability' in the sciences: these concepts and notions can be expected to reoccur in many scientific publications and should be identifiable by way of literature references and 'chains' of citations. Indeed, many papers at a typical conference focus on improvements or refinements of already existing and generally accepted methods and techniques. Hence we can expect the occurrence of 'clusters' of papers from the point in time when a technique or method is first introduced until it is refined to its latest modification. The identification of such 'clusters' thus seems to appear important for the purpose of constructing a HB of high reliability and long term validity. Since conference proceedings are widely available in national or international literature databases, they are not only 'conveniently' accessible, but also, which is methodologically important, open to public scrutiny, verification or critique; this ultimately increases a HB's public 'standing' and trustworthiness.

However, due to the ever growing proliferation of conferences and journals, it must also be determined which ones among those many 'venues' ought to be chosen as the informational basis of 'settled' knowledge for the construction of a HB. Some of these conferences are well established and have been held in series that already span several decades, so that they can be assumed, prima facie, to contain much 'settled' knowledge, while others were started rather recently, but might nonetheless already contain highly relevant papers. In such cases, the difficult to quantify 'reputation' or 'esteem' of a conference (among the experienced experts of the domain's community) might be more important than a conference's quantifiable age, especially if we take into account that 'old' conferences might have considerably 'evolved' their topical focus as the years went by.

In our book's example case, Formal Methods of computing, insofar as they are practically applied in the railway domain, our collection of printed (re)sources spanned almost forty publication years during which these resources had been highly recommended by experts of the domain. However, no 'oral' interviews with specifically chosen practitioners were carried out in our case.

3.2 Step 2: Data Collection and Choice of Domain

Not all information in the knowledge corpus obtained during the previous step can or should make it into a well organised handbook. This is because handbooks only seek to preserve established scientific and engineering knowledge that applies to typical real world problems to be solved in a HB's domain. Thus, it is necessary to filter and select only the relevant knowledge from the large corpus of literature (perhaps even including survey results from informal interviews with capable practitioners). Identifying this particularly relevant knowledge requires significant domain expertise, both 'in general' , as well as in its finer details.

A guideline for this process of filtering is provided by Vincenti's above mentioned six categories of engineering knowledge. Not each and every paper in the chosen corpus might contain knowledge of all six categories, some papers might have a stronger focus on some of those categories than other papers. In this way, the initially single corpus can be usefully divided into several subcorpora ('clusters' of documents within the corpus) along the lines of Vincenti's above mentioned categories. As a result, however, we must also keep in mind that these six categories of knowledge are neither entirely mutually exclusive, nor do they exhaustively comprise all possible types of knowledge [84]. Hence the division of the initially given corpus into subcorpora along the lines of those six knowledge categories is not 'straightforward', such that the same initial corpus might be differently divided into various subcorpora by different experts. For example, it might not always be easy to decide whether some 'knowledge item' found in the literature would fall into the category 'Practical Considerations' or rather in the somewhat related category 'Design Instrumentalities'. In spite of these methodological difficulties, Vincenti's six categories are nonetheless helpful for bringing some form of 'order' into the initially rather amorphous corpus of documents, so that anything which does not 'fit' into any of those six categories at all is, prima facie, (perhaps with some exceptions) a 'candidate' for exclusion from further consideration in the HB's composition process.

Other relevant characteristics of the papers of our corpus must also be identified in this step. For this purpose some guiding questions, such as the following ones, can be asked:

- Does the domain consist of distinct areas or subareas?
- Is it possible that the technical problems being solved in this domain are so varied that different groups ('pockets') of settled knowledge may exist within subdomains of the larger domain?
- How can we 'map' a single paper onto the above mentioned areas and epistemic categories?
- Is the age of a publication a useful indicator of 'settled' knowledge?
- Does the paper contain interesting indicators of 'settled' knowledge that are not 'covered' by the above mentioned categories?

The domain expert and the scientific readers must also agree about terminological issues in this step. They must stipulate precise definitions of the key terms of the domain, must decide which syntactically different terms should be regarded as synonyms of each other, or have other semantic relationships with each other (e.g., specialisation or generalisation). In particular, homonyms (same word with different meanings in different contexts) can lead to dangerous misunderstandings, with the consequence that considerable effort must be put into the detection and clarification of homonyms. These efforts shall ensure, as far as possible, that discussions about and within the given domain's universe of discourse are not obstructed by any semantic ambivalence.

At the end of this exercise, a set of formal 'objects' and formal 'attributes' are identified in preparation for formal concept analysis (FCA). As a result, the 'objects' are the published papers (or other uniquely identifiable sources of relevant knowledge), whereas the 'attributes' refer to the relevant binary yes/no features by which the contents of those papers (objects) can be characterised. For example, object o_{127} could be a paper about railway track control by means of Petri nets, such that *'safety'* as well as *'discrete method'* could be two of the (many other) attributes of o_{127} for which the 'yes' boxes would have to be marked.

At this point, the domain expert and the scientific readers must also collaborate with the FCA expert to determine what other characteristics of the given objects (papers) may be useful in order to automate the identification of settled knowledge. This, however, is not an easy task, as all attributes for FCA must be binary, so that transformation into some suitable binary representation might be needed for attributes that are 'naturally' of multi valued type. For example, a paper's year of publication might be considered as highly relevant information for the issue of settledness; alas we cannot simply define a non-binary *age* attribute with numerical values in the naturals $\mathbb{N}$. To represent a paper's age in binary form (for FCA), the experts could, for example, define three binary yes/no attributes 'recent', 'older', and 'very old', and must then come to some additional domain specific agreement about which one of those three yes boxes ought to be marked in case a given paper (object) is, for example, 4 or 8 or 16 years old (because the community's opinion about what is a 'new' or an 'old' paper can vary considerably from domain to domain). Thus the transformation of 'naturally' multi valued attributes into binary ones (for the purpose of FCA, see below) is a non trivial but, nonetheless, necessary task in this step. The resulting 'objects and attributes' structure will help us to determine (with FCA tool support) how 'settled' the given knowledge really is.

3.3 Step 3: Application of FCA to the Data Collected

Formal concept analysis (FCA) [29] is a branch of 'applied' *lattice theory* [8]. FCA is thus a discrete mathematics based technique of data analysis, knowledge representation and information management by means of which we can identify conceptual structures in data sets [16]. FCA can be used for analysing attribute-object tables (formal contexts) and for exploring the various dependencies which exist between the formal objects and their attributes [90]. The application of FCA to the source data that we have chosen is an important step in our method of composing an engineering HB.

In this theory, a formal *context* $\mathbb{K}$ has a structure $\mathbb{K} = (G, M, I)$ where G and M are sets which represent *objects* and *attributes*.[3] I is the binary *incidence* relation between G and M whereby $I \subseteq G \times M$ and gIm indicates that the object g has the attribute m.[4] For a formal context $\mathbb{K} = (G, M, I)$, operators $\_^{\uparrow} : 2^G \to 2^M$ and $\_^{\downarrow} : 2^M \to 2^G$ are defined for every $A \subseteq G$ and $B \subseteq M$ by

$$A^{\uparrow} = \{m \in M \mid \text{for each } g \in A : gIm\}$$

and

$$B^{\downarrow} = \{g \in B \mid \text{for each } m \in B : gIm\}.$$

Within such a context $\mathbb{K}$, we define a formal *concept* as a pair (A, B) with $A \subseteq G$, $B \subseteq M$, $A = B^{\downarrow}$ and $B = A^{\uparrow}$. A and B are called the *extent*, respectively, the *intent* of the formal concept (A, B).

Explanation: (A, B) is a formal concept if and only if A contains just objects sharing all attributes from B and B contains just attributes shared by all objects from A.

Moreover, the relationship between a *sub*concept and a *super*concept is mathematically characterised by [29](def. 21):

$$(A_1, B_1) \leq (A_2, B_2) \iff A_1 \subseteq A_2 \quad (\iff B_1 \supseteq B_2)$$

The set of all formal concepts of context $\mathbb{K}$, together with their defined order relation, is denoted by $\mathfrak{B}(\mathbb{K})$. The visual representation of this partially ordered structure is called a *concept lattice*.

---

[3] From the German original: K for 'Kontext', G for 'Gegenstände' (objects), and M for 'Merkmale' (features) [29].

[4] An additional remark for the sake of completeness: The 'pure' theory permits the stipulation of non specific, attributeless 'mystery objects', g_0, for which $I(g_0) = \{\}$. If an application domain is characterised by its totality of attributes, M, then all we know about such attributeless 'mystery objects' is that they do not belong in any way to this application domain. Hence they can be practically excluded altogether from any further considerations about this domain.

	needs water to live	lives in water	lives on land	needs chlorophyll	dicotyledon	monocotyledon	can move	has limbs	breast feeds
fish leech	×	×					×		
bream	×	×					×	×	
frog	×	×	×				×	×	
dog	×		×				×	×	×
water weeds	×	×		×		×			
reed	×	×	×	×		×			
bean	×		×	×	×				
corn	×		×	×		×			

Fig. 3.1 *Structure of a Cross Table in FCA: Example from [29](figure 1.1).*

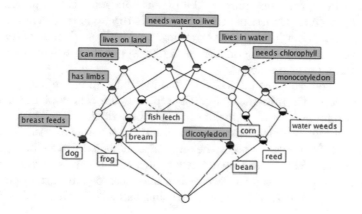

Fig. 3.2 *Concept Lattice: derived from figure 3.1.*

For an object $g \in G$, its object concept $\gamma g := (\{g\}^{\uparrow\downarrow}, \{g\}^{\uparrow})$ is the smallest concept in $\mathfrak{B}(\mathbb{K})$ whose extent contains a g. Additionally, for an attribute $m \in M$, its attribute concept $\mu m := (\{m\}^{\downarrow}, \{m\}^{\downarrow\uparrow})$ is the greatest concept in $\mathfrak{B}(\mathbb{K})$ whose intent contains m.

A formal context can be represented in the form of a Boolean matrix called a *cross table* [29], an example of which is shown in figure 3.1 (adapted from [29](figure 1.1)). In such a table, each row represents an object and each column represents an attribute. A cross (X, meaning 'yes') at the intersection of a row and a column indicates that the object is characterised by that particular attribute.

When such a context-representing cross table is combined with a partial order relation $\leq$ on pairs, we obtain an instance of the above mentioned concept lattice, the abstract type of which is a Galois lattice [29](sect. 0.4). It allows us to 'visualise' the relationships in the given context and, thus, to detect structures and 'clusters' within its data.

The information collected in the previous step of our method must thus be inserted into such a cross table in this step. The FCA expert usually completes this task with the help of a tool that supports the generation of a concept lattice from the formal context provided. Examples of such FCA tools are ToscanaJ, Lattice Miner, Coron, FcaBedrock, or Conexp.[5] The last is the one we have used for [36] [59], as well as for the results described below in chapter 4 of this book.

An example of such a concept lattice is shown in figure 3.2 (automatically derived from figure 3.1 and conceptually identical with [29](figure 1.2)). Concepts closer to the lattice's top node are 'more general' than those below them, which are 'more specific'.[6] A concept at the top of an edge in the lattice graph is called a *parent* concept in relation to the concept at the bottom of that edge, which is called a *child* concept (relative to the parent). If a child concept has more than one parent, the parent concepts share a subset of attributes of the child. Consequently, the lattice's top node represents the least specific concept 'Everything' (or 'Anything'), which has *all* the context's objects in its extent; depending on the particular application scenario, its intent may (or might not) be empty.[7]

Each node in a concept lattice, depicted by 'balls' in figure 3.2, represents a single concept. Certain tools, like Conexp, use the radius of the nodes to represent the number of objects that are members of this concept; the bigger the 'ball', the more objects belong to its extent. In the above example, we can see that all balls appear to be of the same size. This is because, in this case, this visualisation feature had been turned 'off'. In our case study in Chapter 4, however, we are strongly interested in the sizes of our formal concepts' extents, so that we had always switched this ball size feature 'on', see, e.g., figure 4.3.

Moreover: if the visual representation of a lattice node shows a blue filled upper semi circle, there is a so called *own attribute* affiliated with this concept. An own attribute is 'unshared', i.e., it belongs solely to this concept (and, hence, its children). If a node contains a black filled lower semicircle in the visualisation, then there exists at least one object in this concept's extent which is characterised *exactly* by this concept's attributes (no more and no less). Such an object is this concept's 'unshared' *own object*, which does not occur in the extents of any other unrelated concepts.

---

[5] http://conexp.sourceforge.net/ https://www.upriss.org.uk/fca/fcasoftware.html https://fca-tools-bundle.com/

[6] The existence of a lattice's bottom node ($\perp$) is a formal lattice theoretic necessity, which does not, however, always correspond to any positive concept in the application domain; in many typical application scenarios, its extent is empirically empty. The extensional emptiness of $\perp$ is a logical necessity in all those application scenarios in which two attributes $m, \hat{m} \in M$ are mutually exclusive in their 'semantics', so that *no* $g \in G$ can be characterised by both m and $\hat{m}$, e.g., 'rectangular' and 'spherical'.

[7] The top node's intent is nonempty if there exists at least one $m \in M$, which is 'shared' by *all* $g \in G$. Otherwise, the top node's intent is empty, in which case the entire lattice represents a Wittgensteinian *Familienähnlichkeit* ('family resemblance').

Once this process is completed, a formal concept lattice representing the knowledge of our domain can be inspected. In our domain specific scenario, namely Formal Methods and their applications in the railway domain [9], in which our 'objects' are publications and our 'attributes' are the various themes and topics about which those papers were written (see chapter 4 below), it is highly unlikely that we would find a paper that contains a complete description of the entire domain. Hence we expect our domain specific lattice's bottom node ($\perp$) to merely represent a pseudo concept, the extent of which will be empty. Since most papers in the engineering domain are quite topic-specific (i.e., typically describing one main method in comparison with only a few other methods), we can also expect most of our domain-specific paper objects to be merely characterised by a few attributes *each* (though the entire attribute set M for the domain as a whole might be very large). Accordingly, the resulting domain specific lattice will have its own peculiar overall 'shape', which we discuss in chapter 4 below.

Important Note: The property of *settledness* (of domain specific technological knowledge), which is our most important 'epistemic interest' in this phase of the HB composition process, is *not* represented explicitly by any attribute $m \in M$ of our domain; otherwise, we would not need to go through this lengthy process of lattice construction at all! Rather, we must 'infer' this (meta) property from a 'hermeneutic' *interpretation of* the lattice's contents (as described in the subsequent paragraphs). Whereas the (explicit) attributes in our formal concept lattice characterise their corresponding *objects* (i.e., the published papers which we consult as sources of domain specific knowledge), the (meta) property of 'settledness' refers to the *concepts* which are *deduced* by means of FCA *from* those objects and their explicitly denoted attributes. The necessity of some additional interpretive 'hermeneutics' (as described below) stems, thus, from the fact that 'settledness' is *not* an inherent (intrinsic) property of any of the lattice's structural components, as such.

3.4 Step 4: Choice of a Stability Threshold

The lattice constructed in the previous step 3 contains formal concepts ('nodes') in very large numbers, as it grows exponentially in relation to the number of objects and attributes in its underlying cross table. Typically, such a lattice contains far more formal concepts than there were objects in its cross table, so that, usually, many of the very many formal concepts in the lattice will have only a few (or even very few) objects in their extents. As the lattice nodes (formal concepts) in our application scenario are representatives of 'engineering knowledge', we cannot reasonably ignore the question of which of these concepts have 'large' extents (i.e., are 'supported' by many 'witness' papers in our object base) or which ones have only 'small' extents (i.e., are 'supported' by only a few 'witness' objects). Technically, this question, how many or how few 'witness' objects per formal concept in the lattice, is expressed by the notion of a formal concept's *stability*, which is the topic of this section. As 'stability', in this sense, is merely a numeric value, we must finally decide 'reasonably' and 'with the application of expertise' which stability value we want to regard as 'good enough' for subsequent considerations, or which stability values we want to regard as 'too bad', so that their corresponding formal concepts shall be excluded from further consideration in our overall HB construction procedure.

The mathematical definition of '*stability*', together with a corresponding '*stability index*', can be found in [17] [22] [60] and is given as follows: For a context $\mathbb{K} = (G, M, I)$ and a concept $c = (A, B)$,

$$Stab(c) := \frac{|\ \{s \in \wp(Ext(c)) \mid s^{\uparrow} = Int(c)\}\ |}{2^{|Ext(c)|}}$$

This is the relative number of subsets of the concept extent (denoted by $Ext(c)$) whose description (the result of applying $\_^{\uparrow}$) is equal to the concept intent (denoted by $Int(c)$), where $\wp(P)$ is the power set of P.

According to the mathematical characterisation above, stability indicates the *degree of independence of a concept's intent from its extent*. Thus, stability implies some degree of 'noise resistance'; a stable concept does not immediately 'collapse' in case a few objects would be removed from that context — in other words: such a concept would not merge with a different concept, nor disintegrate into smaller concepts, in such a case.

In our domain specific scenario, 'noise' in the knowledge representing lattice (in the form of 'unstable' concepts) should be expected, because the 'objects' of our scenario are papers (publications) and their 'attributes' are related to the various research topics about which those papers were written. Peculiar publications, which are occasionally dedicated to 'unpopular' or rarely discussed 'fringe topics' in the engineering domain, are thus with some likelihood represented by rather unstable concepts in our lattice. Therefore, in contrast to the authors of [22] who are explicitly interested in unstable concepts, we want to 'maintain' only sufficiently stable concepts in our lattice and 'prune' (eliminate) the unstable ones. This desire, however, leads to the methodological problem of how to *choose*, and how to *reasonably justify*, a number $\hat{s}$ such that any stability $s < \hat{s}$ shall be *normatively* regarded as 'too low', whereas any $s' \geq \hat{s}$ shall be regarded as 'high enough'.

Important Note: The lattice itself does *not* contain this information about $\hat{s}$, which is the reason why 'expert wisdom' and some 'hermeneutics of engineering' are needed at this point.

In any case, in this step we extract the most relevant domain specific knowledge by selecting concepts with the 'highest' stability indices, whatever value of $\hat{s}$ we might have chosen for this purpose. Methodologically, we conjecture that this method of eliminating irrelevant information only at this late stage of the HB construction process (i.e., only after the construction of the initial lattice), though it is, admittedly, based on a deliberate choice of a value for $\hat{s}$, is still better (i.e., more justifiable) than a too early pre selection (based merely on 'gut feeling') of those paper objects that shall be fed into the lattice generating cross table in the first place; for further details and examples see chapter 4.

In the theory of FCA, plausible methods for the systematic 'post processing' of lattice data, which can reasonably inform a justifiable choice of $\hat{s}$, have been described and explained in [17] [60] [61] [77] (so that such a 'post processed' lattice is not necessarily any longer *one* coherent data structure [77]). Nonetheless, the possibility of accidentally eliminating 'interesting' and 'relevant' information for the sake of clarity (i.e., over simplification or over abstraction) by means of any such techniques is a risk that the HB editors must 'wisely' take into their methodological and editorial considerations. Therefore, a considerable amount of 'expertise' is needed for the construction and presentation of the 'most appropriate' final lattice, from the concepts of which the table of contents (ToC) of the envisaged HB shall be 'extracted'. This entire procedure thus can*not* be fully automated, though the relevant editorial choices are aided and well informed by automated information processing software tools.

In summary, an appropriately 'balanced' stability threshold value $\hat{s}$ is the result of this 4th step, so that the remaining 'pruned' lattice will only contain sufficiently stable formal concepts that 'best' represent the 'most relevant' domain specific knowledge about which the envisaged HB shall be written.

3.5 Step 5: Classification of Settled Knowledge

The pruned final lattice contains many formal concepts with stability values $s' \geq \hat{s}$, according to the previous step, but it is not yet a reasonably 'linear' table of contents (ToC) of a legible and useful HB: keep in mind that a lattice order is only partial, not 'linear'. Hence further steps are still needed in our HB construction method that 'bridge the gap' between the lattice's many partially ordered concepts and the envisaged HB's linear ToC. As the final lattice, even after its $\hat{s}$-based 'pruning', is still very large, one of the most important questions in this context is *how the concepts in the lattice can be reasonably and 'meaningfully' grouped*, so that a group of concepts can eventually become a 'part' or 'chapter' or 'section' or 'subsection' of the to be written HB.[8]

---

[8] This topic is related to the organisational problems of a librarian who has obtained a new book and must now decide onto which shelf of the library it shall be put; typically, there are several placement choice options which would all 'make sense' when seen from one or another point of view.

The necessary task of meaningfully grouping (or 'clustering' or 'categorisation') the available lattice concepts cannot be blindly 'mechanized', because a machine or an algorithm cannot know, a priori, what 'makes sense' to us or 'what we are after' (with particular 'pedagogical' aims and purposes in mind). On the other hand, such 'clustering' should also not be entirely arbitrary and ad hoc; 'reasonable' principles and a methodologically justifiable 'hermeneutics of engineering' are thus needed in this step, too, though, again, we will not disregard the opportunities of automated or semi automated decision support. Helpful algorithms for classification or clustering are well known [23] [25], and we recommend their application, as long as a human domain expert will still be there to review, and possibly improve, the 'suggestions' that those algorithms yield as their outputs; these algorithms might suggest 'strange' or 'artificial' groupings which would have no 'canonical match' in the internal organisation of an engineering domain with which the engineers (practitioners) are already familiar. A purely manual classification attempt (without any algorithmic tool-support), on the other hand, would be a daunting (if not infeasible) task if the underlying concept lattice (even the 'pruned' one from the previous step) is very large.[9]

Thus, all in all, the domain expert evaluates in this 5th step whether all the previous (tool supported) steps led to a satisfactory result, i.e., if the 'subjective' view of the domain expert and the 'objective' analysis results are 'well aligned' with each other. If, at this point, the overall result appears to be unsatisfactory to the expert, several possible causes must be taken into account:

- the initial corpus of domain specific literature, G, might have been poorly chosen (relevant sources omitted or irrelevant sources inserted);
- the set of attributes M for the initial cross table might have been inappropriately defined (with too many or too few $m \in M$);
- for any source paper $g \in G$, its attribute matching incidence relation, $I(g) := M_g \subseteq M$, might have been wrongly stipulated (with relevant attributes omitted or with irrelevant attributes assigned);
- the stability threshold value $\hat{s}$ (defined in order to 'prune' the initial lattice) might have been inappropriately chosen;
- the clustering algorithm, which was chosen out of many possible algorithms in order to group the final lattice's formal concepts into larger categories, was not a suitable algorithm for this particular domain specific task, or might been been wrongly applied with inappropriate initial parameters, etc.

In any of such cases, any one of the aforementioned steps (1–5) might have to be redone until the experts are satisfied with the final result. As a consequence, as usual in all engineering projects, the 'early' errors, if detected too late, are much more 'expensive' in relation to their 'correction costs' than

---

[9] A convincing application example of such FCA and lattice based semi automated classification methods in the domain of software re-engineering can be found in [7].

the errors made at later stages. The 'editorial engineers' who are responsible for an envisaged HB composition project, must also, like any other engineers and project managers, keep these cost threats in mind.

3.6 Step 6: Presentation of Settled Knowledge

Finally, the presentation of 'settled knowledge' for inclusion into a HB should follow a standardised presentation scheme which reasonably reflects the typical style and characteristics of 'normal' engineering methods [84], particularly in that domain. The envisaged HB will contain many entries, so that, to avoid confusion, some 'uniformity' in the presentation of all these entries ought to be maintained. In particular, the content of each HB entry should reflect its associated 'category of engineering knowledge' according to Vincenti; see the list of categories at the beginning of this chapter. Other (similar) schemata or categories of knowledge might also be suitable and appropriate from case to case.

In our HB construction method, we advocate the following *problem oriented* approach to a well structured presentation of HB entries. In other words, each HB entry shall be given according to the following presentation scheme; for a concrete example, see chapter 5.

Problem Class:
   Characterises generally the types of problems for which solutions are described and explained by this HB entry (possibly with subclasses).
Solutions:
   Describes the family of possible and appropriate class specific solutions, one after the other. To this end we also provide the following additional helpful information together with each solution:

Criteria:
   Here we describe the prerequisites and conditions under which a potential solution is actually applicable.
Principles:
   Here we explain briefly on which laws or scientific theorems a proposed solution is grounded. This is necessary because engineering is a science based activity which 'borrows' its trustworthiness from the truth of the underlying scientific assertions; for comparison see [14](sect. 11.2).
   • For example, there are different physical principles which can be the basis for a solution to the problem of constructing a temperature measuring device; these could be 'liquid in tube' or 'bimetallic strip'. In computer science, computational problems can be addressed by different problem solving approaches, such as greedy algorithms, divide and conquer algorithms, backtracking algorithms.

Process:

Here we briefly outline, step by step, the method for producing the desired solution to the problem at hand. Depending on the 'size' of the problem scenario, such a procedure might involve several analysis steps, design steps, manufacturing or implementation steps, and the like. A full elaboration of these steps and their possibly many substeps, however, cannot be given at this point; literature references to relevant textbooks (e.g., software development management or project management), in which many further details can be found, must suffice at this point.

Validation:

Here we outline briefly the techniques of proof (mathematical, theoretical) or testing (experimental, empirical) that ought to be applied in order to demonstrate with sufficient plausibility that a designed solution (candidate) does indeed solve the problem that it was meant to solve, and that this solution also does not accidentally create unacceptable new problems (dangerous situations or otherwise harmful consequences) as unintended side effects.

Further Reading (optional):

In cases where a HB entry's problem class and solution proposals are non trivially 'large' or 'complex', so that they cannot naively be assumed to be 'well known' amongst all practitioners in its field, a reasonable number of literature references (not too few and not to many) might provide helpful pointers to further details which, for the sake of brevity, the HB entry itself cannot provide.

At this point the HB editors and authors must not forget that, in the first place, a HB ought to address the *practitioners* in the field (not the academics and researchers), which implies that there should be no exceptional 'ingenuity' required to follow the problem solving process that a HB entry describes. In other words: *no* substantial 'innovations' or new 'inventions' (at a significant level of originality) will be needed at such a point, though it is (of course) presumed that a seasoned practitioner, who reads such a HB entry, will be competent and experienced enough to come up with small, case specific 'modifications' of his own; this is what the earlier use of the word 'improvement' intended to convey.[10]

In the next chapter 4, we present a domain specific example, Formal Methods of computing in the railway domain, of the HB construction steps 1–5 that we have motivated and explained in the previous sections of this chapter. An example of our method's 6th step follows thereafter in chapter 5.

---

[10] An illustrative example can be found in the system of [80], which provides software developers with help for the derivation of correct and efficient programs from formal specifications.

Chapter 4
Application of the General Method to the Railway Domain

What knowledge of Formal Methods in the railway domain is already 'mature'? To answer this question we must consider the most suitable sources of knowledge in our chosen domain, as well as a suitable definition of the notion of 'settledness'; see chapter 3. To this end, it is also helpful to investigate how the 'discovered' knowledge is *internally structured*, so that it can be properly classified according to our 'epistemic needs'. The answers to these questions will lead us to some appropriate *attributes* that will be included in a *formal concept lattice*. This lattice will represent the 'essence' of the gathered knowledge. After the construction of an initial lattice, which still contains some level of 'noise' (i.e., information of only little practical value), its 'pruning' by way of so called *stability indices* leads to a final lattice which shows the most 'stable', hence, 'settled' knowledge of practical usability in the railway domain. As mentioned above in chapter 3, the quality or 'practical value' of this lattice, as a whole, depends strongly on the quality of the chosen sources from which it was extracted and abstracted. This final lattice is then 'interpreted' by the experts in order to obtain a classification, so as to organise the relevant knowledge in a useful manner. Finally, this organisation will be 'materialised' in the structure of the envisaged HB's table of contents (ToC).

Formal Methods of computing are almost always applied under somewhat 'simplified' ('idealised' or 'generalised') circumstances, at some reasonably high level of abstraction. This is only partly due to high computational 'costs';[1] these 'generalisations' (at higher levels of abstraction) also make those techniques widely (re)usable in various circumstances and, hence, suitable for 'HB-ification'. Because the level of conceptual abstraction in formal models is typically high, *refinements* or variations (depending on the overall aim of the (software) engineering project in which those formal models occur) of those models belong to the typical tasks in many projects in which Formal Methods are applied. As mentioned above, all this is also true for the

---

[1] Many problems in this field, such as the satisfiability problem, are NP-complete.

© The Author(s), under exclusive license to Springer Nature Switzerland AG 2020
S. Gruner et al., *On the Construction of Engineering Handbooks*, SpringerBriefs in Computer Science, https://doi.org/10.1007/978-3-030-44648-2_4

railway domain [26]. Identifying the generally applicable Formal Methods (at a 'broad scale', without their finer details) must thus be the first step towards composing a HB on this topic. As mentioned above in chapter 3, this task can be solved reasonably well by means of literature search (*unless* relevant practitioners in the industry would apply secret 'in house' methods about which little is known in the public domain).

4.1 Step 1: Choice of Sources for the Railway Domain

The *general* knowledge sources for HB composition projects were already listed and explained above in section 3.1, so that we do not need to repeat them at this point. In our specific example case [36] [59], by means of which we illustrate our HB construction method throughout this book, only published conference papers were taken into account.[2] The papers chosen in our case span more than three decades of domain specific scientific activities and were taken from the following conferences, which belong to the most important meetings of experts in the field of software supported railway engineering:

- Proceedings IFAC: 1975-2012,
- Proceedings FORMS-FORMAT: 2010-2014,
- Proceedings SAFECOMP: 2005-2014.

Altogether, these proceedings provided more than 300 potentially relevant literature sources. However, about 150 of these papers did not relate specifically to the use of Formal Methods and were thus excluded from our data base [59]. Of course, many more railway related Formal Methods papers can be found in various other repositories.

4.2 Step 2: Data Collection and Choice of Domain Specific Attributes for FCA

In our method of HB construction, the choice of *formal concept analysis* [29] leads to an attribute based classification system of domain specific knowledge. In our example case of knowledge about Formal Methods and their application, the simplest possible classification scheme might merely combine the 'name' of the Formal Method used in each literature source together with its 'age' (w.r.t. to the question of 'settledness').

---

[2] At this point the reader ought to remember that we are presenting a HB construction *method* in this book at the 'meta' level — *not* a complete 'body of knowledge' of all Formal Methods in the railway domain at the 'object' level. Therefore it was not necessary for this purpose to seek and exploit all the possible sources of knowledge which section 3.1 has listed and explained.

To judge whether a paper had been appropriately included in our data base, the 'scanning' of the *keywords* at the beginning of each paper was deemed as a good starting point, at least for the purposes of illustration in this book. However, as different Formal Methods are used in different contexts for solving different problems or classes of problems in the chosen application domain, additional keywords (to capture more context sensitive information) were declared as 'refining' attributes in our classification scheme. Whilst a more comprehensive explication of those attributes can be found in [59], the following points should suffice to illustrate the principles of our approach:

- In addition to 'when' (year), it was also necessary to know 'where' in the railway domain those Formal Methods were used. To this end we recorded the *subdomains* of the railway domain to which each publication pointed. The following subdomains were explicitly represented in our classification scheme: the 'Net', 'Timetables', 'Scheduling and Allocation', 'Traffic Monitoring and Control', 'Rolling Stock', 'Passenger Handling', and 'Freight Handling'.[3] Hence the more 'safety critical' a subdomain is, the more applications of Formal Methods we should be able to find in it.
- Further attributes can be included into the classification scheme in order to also represent the 'scientific maturity' levels of the analysed source publications. Shaw's maturity grading scheme, which is suitable for this purpose, can be found in [78]. Scientific maturity in this context is indicated by the theoretical 'depth' of mathematical, descriptive or analytical techniques, as well as by the precision and reliable repeatability of experimental evaluations in support of theoretical hypotheses [78]. Hence any literature source characterised by a *low* level of scientific maturity should perhaps *not* be considered as a likely candidate for 'HB-ification'.

4.3 Step 3: Application of FCA to the Data Collected

According to [29], a 'formal context', which is a set of 'objects' together with their descriptions in the form of 'attributes', generates a family of 'formal concepts'. Hence each of these many formal concepts has an 'intent' and an 'extent'.[4] The extent of a concept consists of all (formal) objects which belong to this concept, whereas the intent of a concept consists of all the Boolean attributes that apply to all the formal objects of that concept [29]. In our approach, the objects are the sources of knowledge in the form of the published papers from the above-mentioned repositories; each paper is one 'object'. The attributes stem from our 'empirically' observed features of those papers, i.e., when they were published, which subdomain they belong to, which Formal Methods they advocated, etc. With these objects and attributes, a large FCA

---

[3] http://euler.fd.cvut.cz/railwaydomain/

[4] Synonyms are 'intension' and 'extension', respectively.

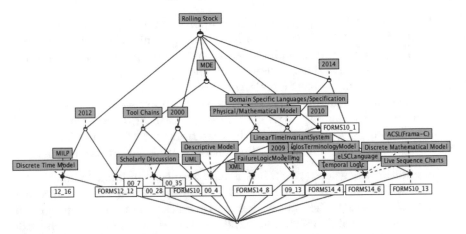

Fig. 4.1 *Sublattice representing Knowledge in the Rolling Stock Subdomain.*

cross table for the railway domain was constructed: see [59] for all its details. From this large cross table (too big for a legible and helpful graphical reproduction in this book), fed as input to the ConExp tool, we were able to generate the corresponding Formal concept *lattice*. Its *partial order* represents the various (binary) relations between the included concepts; some are mutually exclusive (disjoint), others are sub or superconcepts of each other, some are 'overlapping' in sharing a common superconcept, etc.

Concept lattices grow exponentially in relation to the numbers of objects and attributes [62], so that the insertion of even only a modest number of objects (in our case, papers) will quickly lead to 'unwieldy' lattices which demand large computational resources for their automated processing. *Human expertise* is thus still needed in our method to 'feed' the FCA's input data structure (i.e., the cross table) sufficiently 'wisely', so as to avoid the notorious 'state space explosion' problem.[5] In our case study, the resulting lattice for the entire domain was so large and complex in its internal graph structure that it would not be illustrative to depict it graphically here in this book.

For this reason, figure 4.1 shows only our sublattice of the previously identified subdomain 'Rolling Stock' [59]. In the structure, 'model driven engineering' (MDE) can be found as a prominent software development technique for which Formal Methods are relevant. In the same figure, we can also see that the sources of that knowledge stem from the years 2000–2014; thus, they have a reasonably long historic duration, which might be regarded as indicative of a reasonably high level of maturity or 'settledness'.

---

[5] For this reason we recommend the separate application of our method for each a priori identified subdomain (yielding a smaller lattice per subdomain) rather than for the entire domain (yielding a no longer feasible, huge lattice for 'everything').

These smaller subdomain specific lattices are not only computationally more 'feasible'; they also permit a reasonably human friendly 'visualisation' of the given data set so that its relevant conceptual relations are easily observable by the 'naked' human eye.

At this point, our concept lattice(s) represent(s) *all* the domain knowledge (including the 'noise') extracted from our chosen sources of information, but we still need to determine which parts of that information may be considered as 'matured' or 'settled' enough for our purpose in constructing a HB. The 'noise' in our lattice (which represents somewhat irrelevant concepts, including, in our case, 'immature' or not 'settled' knowledge) must still be filtered out for the sake of more accurate and more relevant analysis results. For example, the lattice might include an insignificantly small number of 'objects' (papers) in which some formal techniques for the railway domain were merely 'proposed' or 'advocated' or 'attempted' without much subsequent 'echo' over longer periods of time.

4.4 Step 4: Choice of a Stability Threshold

In the formal concept lattices corresponding to our chosen subdomains of the railway domain, we can find many concepts which have *merely one object* in their extents. This observation is not specific to the railway domain, but is rather characteristic of our approach in which we observe a domain over an extended period of publication time. Understandably, the more suitable and more successful Formal Methods are applied and described more frequently in their application domains' publications than those with single references, which turn out to be rather disappointing. A typical example is included in figure 4.1: the extent that captures the use of the Frama-C language (by means of which one can discretely model Live Sequence Charts) in the 'Rolling Stock' subdomain is of size 1, i.e., there is only one paper in which this particular Formal Method was advocated. Our data set indeed includes many formal concepts (object-attribute pairs) that appear only once in the lattice. This was to be expected, since exploratory research requires scientists or engineers to propose and test new methods in search for progress and improvements. The failing ones, however, do not reappear in the literature as frequently as the more successful ones.

This 'line of reasoning', which is mainly based on counting, seems, prima facie, to be plausible. However, without deeper and farther reaching methods of analysis, there is no sure way of knowing whether the publication sources (papers) corresponding to those 'singleton' objects are pertinent to the overall knowledge structure, or whether they really are merely 'noise': see [22] for comparison. In our approach those 'singleton' objects are considered irrelevant to begin with, with the methodical consequence that their formal concepts are removed from the final knowledge representing lattice [59]. This decision, however, directly leads to a notorious methodological problem: if only 1 object per extent is regarded as 'not enough', then how many n are still 'too few', and how many n' are already 'a multitude'?[6] We solve this methodological problem by the following 'stability' considerations.

To obtain a suitable *stability threshold* value, we plot a graph of the percentage of data included in the lattice versus their corresponding stability threshold indices (where the underlying definition of the notion of 'stability' is taken from the canonical FCA literature). The resulting *stability variation diagram*, which can be seen in figure 4.2, shows the percentage of that data which would remain in the given lattice if a particular stability index would be used as a 'cut off limit' (threshold) for the purpose of 'pruning' the lattice structure of 'insignificant' formal concepts with 'too small' extents. In our specific example (figure 4.2), it is easy to see that there are two intuitively large 'drops' in the graph of information inclusion: one at 25%, and another one at 50%. This phenomenon is, on the one hand, a consequence of the mathematical equation which defines 'stability', but, on the other hand, also due to the empirically given number of concepts which occur in this specific lattice with the methodologically 'reasonable' amount of at least 3 objects in their extents.

For obvious reasons, we did not choose the stability index of merely 25% corresponding to the first 'drop down' in figure 4.2. Above the second 'drop down' threshold of 50% in the bar chart, the inclusion rates are 'sinking' gradually, so that we reasonably chose the 50% value as our threshold for our above mentioned purposes: between the values of 25% and 50% the percentage of extensional inclusion is nearly constant. Selecting this stability threshold and correspondingly removing the 'noise' from the original lattice yielded the final lattice (for our above-mentioned subdomain), shown in figure 4.3; the explanations of many further details can be found in [59].

Thus, after selecting the most suitable threshold, as motivated above, we 'prune' (remove) from the initial lattice the concepts that fall below the chosen threshold. Consequently we obtain a 'clean' and 'stable' lattice, the formal concepts of which reasonably represent the interesting information structures within our domain of discourse. The Formal Methods represented by this final lattice may thus be regarded reliably as instances of 'settled knowledge' in our domain and, therefore, as 'candidates' for inclusion into the envisioned domain specific HB. In our example case study, they included

- Mathematical Models (in general);
- Discrete Mathematical Models;
- Discrete Event Systems;
- Markov Models;
- Petri Nets;
- Fuzzy Logic.

---

[6] For comparison, see the *sorites paradox* which was already known to the philosophers of Greek antiquity.

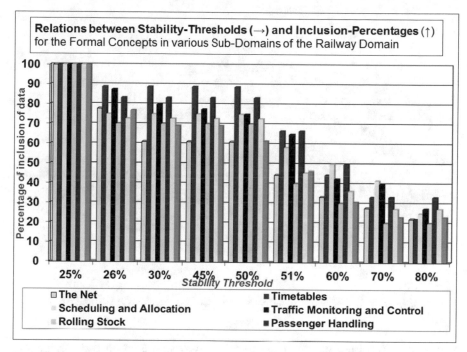

Fig. 4.2 *Changes in data inclusion as a function of the stability thresholds.*

Interestingly, these findings were also mentioned in other, independently conducted, overview studies [26] [34] by means of which we can further 'gauge' the 'methodical value' of our FCA based HB construction approach.

4.5 Step 5: Interpretation and Discussion of the Detected Settled Knowledge

On the basis of figure 4.3, we now need to go through some further steps of 'interpretation' in order to obtain an initial 'classification' of this knowledge. This will help us to organise the knowledge from the lattice in a useful manner, so that engineers will be able to utilise this knowledge effectively in their daily work.

4.5.1 Observations (and Peculiarities)

In our case study [36] [59], we have observed that almost all the modelling approaches and formalisms are related to the 'Traffic Monitoring and Control'

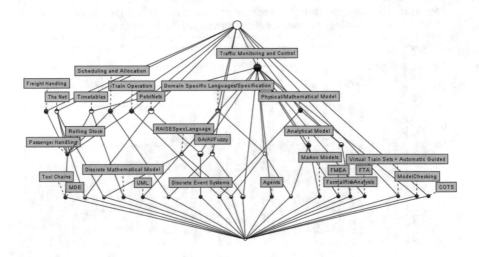

Fig. 4.3 *Final lattice of stable concepts with a stability threshold ≥ 0.5 and extent > 1.*

subdomain, (and not so much to the other subdomains, with the possible exception of 'Timetables'). The modelling languages and formalisms appearing in our lattices find their applications mostly in the design and implementation phases of project life cycles, and to some extent also in the testing phase.

Note also that it was worth noticing that Discrete Event Systems appeared as a more specific concept in relation to the Petri nets concepts, which is not entirely consistent with our previous knowledge. The reason was apparently that the lexical *keyword* 'Discrete Event Systems' was most often used in conjunction with 'Petri nets' in our underlying data base of conference publications. In general, however, there are more mentions of the term 'Petri nets' than of the broader term 'Discrete Event systems'. These findings reveal the influence of 'keyword capturing' (in the early stages of our HB construction method) on the 'validity' of the eventually generated formal concept lattices, and caution us to capture keywords 'wisely' (with expertise) before constructing our approach's initial cross tables for FCA. Indeed, a closer look into this point revealed that every paper on Petri nets somehow discusses Discrete Event Systems as well, even if only as somehow 'related work'. Not taking these finer details into account can thus possibly lead to 'distorted' formal concept lattices in which the generalisation hierarchy (of subconcepts and superconcepts) does not make much real world sense.

Interestingly, our final formal concept lattice also contained (per automatic inference) some nameless (but nevertheless significant) *combination* concepts. For example, we have found in the lattices of [59] a conceptual connection between 'Domain Specific Languages', the language 'RAISE', 'Analytical Models', and the subdomain of 'Traffic Monitoring and Control', though *none* of our many *individual* source objects (published papers) mentioned this 'cluster' of conceptual connections. This observation reveals the ability of our FCA based semi automatic method to '*see more*' than any individual paper author or domain expert alone can see. For the composition of a forthcoming HB on the application of Formal Methods in the railway domain, those automatically detected conceptual combinations may be particularly fruitful, because a HB ought to put every listed method of engineering into its proper context of application.

4.5.2 Settledness

Examining figure 4.3 for examples of settled knowledge, we can find them only under 'Traffic Monitoring and Control'. Indeed, in this field, many attempts at using Formal Methods are known. 'Timetables' and 'Scheduling and Allocation' are not significantly represented in our case study, though they are intensely researched in *other* computer science communities and 'venues', particularly artificial intelligence (AI) and operational research (OR), which were *not* among our chosen conference topics (listed above in section 4.1). These two well known areas indeed contain much 'settled knowledge', without which no account of settled knowledge of Formal Methods in the railway domain would be complete, though our chosen conference 'venues' did not sufficiently 'reflect' this fact. Possible and actual *knowledge transfer* between apparently 'unrelated' fields of research and engineering must thus not be forgotten when a HB for one particular field is to be composed.

4.5.3 Limitations of our Findings

Many of the subdomains of railway engineering in our case study [36] [59] have limited Formal Methods related to them. Hence the discussion of 'settledness' in these subdomains might perhaps be premature. In our case study, there was not enough initial data to support a more 'meaningful' analysis and interpretation. In our case, perhaps the observation period of publication time was too short, or not enough literature was looked at, or possibilities of knowledge transfer from other domains were not sufficiently considered; or, perhaps, there is indeed not yet much 'settled' knowledge in these areas. Similarly, and for similar possible reasons, the identified Formal Methods for the subdomain of 'Train Management and Control' fail to exhibit any deeper substructures for further classification (subconcepts, superconcepts, meaningful 'clusters' of concepts, and the like).

4.5.4 Guidance for a Handbook on FMs for the Railway Domain

On the basis of our case study's findings and their above mentioned limitations, we can currently recommend and advocate the (future) production of a HB of 'settled' engineering knowledge in Formal Methods of computing for 'Train Management and Control', (perhaps even with the addition of 'Timetables' and 'Scheduling and Allocation'). Due to its above mentioned limitations, we cannot use our case study in support of any other recommendations, though other experts might already be in possession of related 'settled knowledge', which our case study was not able to reveal. In any case, as our previous elaborations have shown, any such identifications and classifications need to be carried out very carefully, lest the chosen data base might be too small to support any reliable assertions and conclusions.

How a specific 'piece' of such settled knowledge ought to be 'cast' into a HB compatible *form* is the topic of chapter 5, which follows after the next (and final) section of this chapter.

4.6 Possible Threats to Validity: Critical Evaluation of Steps 1–5 in the Context of our Railway Example

This final section of chapter 4 serves as a continuation of the 'Limitations of our Findings' subsection of the previous section (4.5.3), albeit now on the broader scale of our entire HB construction method (as opposed to merely its final step).

By means of our comprehensive *example* of domain analysis (with further details available in [36] [59]), we have *illustrated* how our (meta) method for the composition of an engineering HB is meant to be carried out. To apply our method properly for the composition of a 'real' HB for any chosen application domain, it is important not to forget any details of relevance, wherein prima facie 'irrelevant' details might well turn out to be crucial in later stages of such a project, lest the resulting HB be incomplete, even if the method of its construction was properly followed.

In our above mentioned illustrations and examples of domain analysis, results were obtained on the basis of several preconditions and 'rational assumptions'. Though all those assumptions can be methodologically defended (see above) as 'reasonable', we must nonetheless make those preconditions and assumptions explicit, so as to avoid any naive methodological over confidence and, hence, the composition of 'incomplete' or otherwise unreliable engineering HBs. For this reason, at least the following *methodologically critical* points must be particularly mentioned.

4.6.1 Notion of 'Settledness'

In asking which knowledge is sufficiently 'settled' for its 'HB-ification', we defined a notion of 'settledness' which is essentially *temporal*. Accordingly, such knowledge must occur and re-occur sufficiently often over a sufficiently long period of time.

- Critical epistemologists might thus ask whether or not our time based notion of 'settledness' is appropriate, particularly w.r.t. Shaw's notion of 'scientific maturity' [78]. In our examples above, we also did *not* consider chains of citations or cross references within the scrutinised body of literature in order to investigate 'settledness'. Moreover: where too much emphasis is placed on the 'history' of knowledge, there might arise a choice bias against 'newer' knowledge, which might nonetheless be accepted and adopted quickly by a community of expert practitioners. For example: in our case study lattice below, the B method does not (yet) appear at all, although the B method is already widely regarded as a 'paradigm' of the industrial applications of Formal Methods in the railway domain. Thus, if the 'settledness' criterion are inappropriately defined, de facto relevant settled knowledge might not be found by the subsequent application of our FCA based methods of domain analysis.

4.6.2 Choice of Database

We have sought settled knowledge in *public* data bases, particularly in the *community relevant* conference proceedings indicated in section 4.1, out of which we extracted a *representative selection*.

- Critical experts might thus argue against us that those chosen conferences were perhaps not 'community relevant'; that the set of chosen conferences was too small (and thus not representative), so that relevant settled knowledge might have been omitted; that knowledge transfer possibilities from other (prima facie 'unrelated') fields and research communities have been ignored; or that the settled knowledge of the railway industry might perhaps exist only in the form of corporate secrets (instead of being publicly available). For example: in the data base of the above mentioned illustration of our construction method, we did *not* include the proceedings of 'smaller' events like the FMERail workshop series, nor journals (rather than conferences), nor 'isolated' papers like [38] or [88] (which had been published in 'general' conferences outside the topic specific meetings of the FM Rail 'community').

4.6.3 Choice of Formal Concept Analysis

We have chosen FCA as our method of domain analysis because FCA has already demonstrated its epistemic usefulness [62] in various other application domains, see, for example, the annual proceedings of the International Conference on Formal Concept Analysis (ICFCA), and may thus be regarded as a trustworthy method in the field of formal epistemology.

- Nevertheless there might be other, different methods that could and should be applied in order to extract initial sources of settled knowledge. Any suitable alternative 'data mining' approaches could or should also be applied in order to confirm or to correct and improve our preliminary findings.

4.6.4 Choice of Attributes for FCA

As mentioned above, we have used the attribute based method of FCA to identify those concepts which we are strongly associated with 'settled' knowledge; see section 4.2. Our declaration of the relevant attributes to be used in FCA's cross tables (figure 3.1) was made 'hermeneutically', i.e., as in the faculty of Humanities, after thorough reading and interpretations of relevant engineering literature.

- Critical experts might thus argue that our chosen attributes were perhaps not identified appropriately, or that we might have wrongly omitted important attributes altogether; consequently, the automatically generated lattice graphs (figure 4.1) would not represent an accurate 'image' of the chosen domain or subdomain. For example, whereas we have taken *subdomains* of the railway domain into account, we did *not* refine our analysis to the finer level of typical *problem classes* within those subdomains.

4.6.5 Choice of Stability Threshold

After a first 'raw' lattice with too many 'unstable' concepts had been obtained in the first phase of our application of FCA, we have selected and included only those concepts with a 'stability' *above* a particular percentage threshold ($0 < \hat{s} < 1$). Though the value $\hat{s}$ was chosen carefully on the basis of 'reasonable' considerations (figure 4.2), it was nonetheless a deliberate *choice*.

- Critical experts might thus argue that our choice of $\hat{s}$ was not appropriate, that an alternative threshold $\hat{s}'$ (with $\hat{s}' \neq \hat{s}$) should have been chosen instead of $\hat{s}$, or that the considerations of [22] (concerning the significance of low stability concepts) ought to have been taken into account.

4.6.6 Classification of Settled Knowledge

In order to obtain a classification of settled knowledge in the railway domain, we analysed the concept lattice and gave it a meaningful 'interpretation'. Thus, we related our findings to our already pre existing 'view of the domain', our 'expectations', and our previous 'knowledge of the field'. This interpretation was a matter of 'hermeneutics', similar to the scholarly methods applied in the faculty of humanities. This interpretation led us to the 'insight' that only selected subdomains of railway engineering seem to lend themselves, at this point in time, to the construction of a HB on this topic.

- Critical experts might thus argue that, in the end, our 'view of the domain', 'expectations', or 'knowledge of the field' played too large a role in the overall analysis, so that the results of our interpretations are not sufficiently 'objective'.[7]

---

[7] At that point we might reply that the way we utilize our 'knowledge of the field' was merely to avoid over interpretation of the data; we did not add any 'artificially invented' concepts, but rather pointed out 'anomalies' and shortcomings of the automatically extracted findings.

Part III
Synthesis

Chapter 5
Example HB Entry of a Formal Method for the Railway Domain — Step 6

In the previous chapter we identified settled knowledge concerning the use of Formal Methods within the railway domain. Now we take a first tentative step from analysis towards synthesis. To this end, we apply the presented meta method to describe how settled knowledge could be presented in a HB. We illustrate our suggestion by providing a sketch for one concrete example, namely, on how model checking can be utilised for safety verification of interlockings written in Ladder Logic.

Our presentation scheme has been developed based on discussions at a workshop affiliated with the International Conference on Software Engineering and Formal Methods in 2013 (SEFM'13) on the topic of a BoK in railway control [35]. It is deliberate that our scheme looks 'familiar', 'natural', 'non surprising'. Its purpose is to be intuitive, easy to understand, allowing the reader to focus on the content rather than on the framework. To achieve this, we took some inspiration from [1]. Naturally, such a scheme could, as convincing as it might look at first sight, turn out to be unsuitable for capturing settled knowledge. Thus, as a proof of concept, in the following we put it to the test by applying it to a well established example in the railway domain, drawing on the verification approach described in [47].

Using software model-checking of Ladder Logic in order to verify safety properties of interlocking programs is well understood in academia. It is generally agreed that it belongs to settled knowledge within the railway domain. For various academic publications see [28] [32] [46] [47] [50] [51] [52] [53] [54] [64] [85] [91], and —for the industrial view on it— an experience report about verification practice [24].

In section 5.3.1, we provide a short bibliography. The discussed method belongs to the 'Traffic Monitoring and Control' area, for which we identified settled knowledge in section 4.5. The corpus that we considered for our concept analysis, however, does not include the publications quoted above.

In the following we present the example and also accompany it with an academic explanation which would not normally be included in a HB. This part is clearly separated from the example HB entry. Sections 5.1–5.2 present

© The Author(s), under exclusive license to Springer Nature Switzerland AG 2020
S. Gruner et al., *On the Construction of Engineering Handbooks*, SpringerBriefs
in Computer Science, https://doi.org/10.1007/978-3-030-44648-2_5

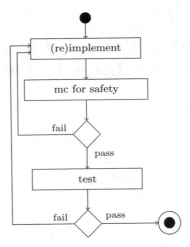

Fig. 5.1 *Process of model checking (mc) for Safety.*

the sample HB entry. Academic explanations can be found in section 5.3. Within the sample entry given in sections 5.1–5.2, we apply the following writing style:

- Cross references from our sample entry to other parts of the still to be written Handbook of Formal Methods for the Rail Domain (HB-FM-RD) are written as, e.g., "section ... on ... in this HB-FM-RD".
- In section 5.2, we illustrate (Step 3) and (Step 4) with just *one* variant. We provide the headlines of other, potential variants in order to illustrate the entry's 'architecture'. However, we do not provide text for these variants as it would structurally be similar to the one already given. We put "[...]" as placeholders for those variants that are not provided.

We conclude this chapter with some experience report on writing a handbook following the scheme together with some readers' reactions.

5.1 Problem Class: Verification through Model Checking of Ladder Logic Programs for Safety

The programming language Ladder Logic is often utilized to program railway interlocking computers. Safety properties for interlockings arise from regulators or customers. Model checking allows for a full system analysis and investigates whether all 'runs' of the interlocking (program) are safe. Model checking is just one verification technique that can be applied. References to other techniques for the same problem can be found in sections ... of this HB-FM-RD.

Inputs

The methods described below require as inputs:

- A ladder logic program accompanied by a variable naming scheme;
- A safety formula in a discrete time, temporal first order logic expressing a safety property; and
- A trackplan with a corresponding route table.

Subclasses

Depending on the user's expectation about the correctness of the program, there are two cases to be distinguished: in the first case, *Expected-Fail*, the user expects verification to fail, i.e., the user's interest is in gaining informative counterexample traces for the purpose of debugging; in the second case, *Expected-Success*, the user wants to achieve a positive verification result.

Alignment with a Verification Step

The UML activity diagram shown in figure 5.1 illustrates how model checking for safety can precede standard (as defined in the railway industry) verification through testing. Note that the final verification step is still testing, i.e., adding model checking does not change the way the safety case is made to the regulator for the interlocking computer. In the present state of industry and regulators, testing would still be required to be done.

Gains

Adding model checking to the program analysis will lead to a higher level of safety assurance as *all* instances of the safety property are considered, rather than just those *selected* instances which are encoded in a test case. It has been demonstrated that certain test approaches are, in principle, not able to uncover mistakes that can be found through model checking. As automated model checking is cheaper than manual testing, adding model checking has the potential to reduce cost.

5.2 Solution

The method has four main steps. The first two of these pre-process the input data; here, there is no variation. The third applies model checking to pre-processed data; here there are several subclasses. Finally, the results need to

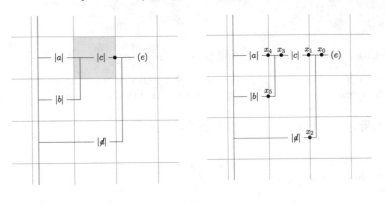

(i) A sample rung   (ii) Illustration of Tseitin Transformation

Fig. 5.2 *An Example taken from Kanso [52].*

be interpreted; this interpretation depends upon the chosen model checking approach.

(Step 1) Translating Ladder Logic into Propositional Logic

Translate the Ladder Logic program into a propositional formula Ψ using the Tseitin transformation [82]: see section ... of this HB-FM-RD. This results in a Ladder Logic formula Ψ in Propositional Logic.

Example (Deriving Boolean Expressions) Reading off the Boolean expression from the Ladder Logic program shown in figure 5.2 (i) in a naive way, one obtains

$$e := \neg d \vee (c \wedge (a \vee b)).$$

Reading it off via the Tseitin transformation introduces intermediate rungs and variables, see figure 5.2 (ii). Concretely, we obtain the following program written as a series of assignments:

$$x_5 := b; x_4 := a; x_3 := x_4 \vee x_5; x_1 := c \wedge x_3;$$
$$x_2 := \neg d; x_0 := x_1 \vee x_2; e := x_0$$

Note that there are more assignments thanks to the additional variables introduce by the Tseitin transformation.   □

The representation of a Ladder Logic program in Propositional Logic is constructed in terms of disjoint finite sets I and C of input and output variables, where internal variables are subsumed in C. We define $C' = \{c' \mid c \in C\}$ to be a set of new variables (intended to denote the output variables computed in the current cycle). In addition, we need a function unprime $: C' \to C$, unprime$(c') = c$.

Definition (Ladder Logic Formulae) A Ladder Logic formula ψ is a Propositional formula

$$\psi \equiv (c'_1 \leftrightarrow \psi_1) \wedge (c'_2 \leftrightarrow \psi_2) \wedge \cdots \wedge (c'_n \leftrightarrow \psi_n)$$

such that the following holds for all $i, j \in \{1, \ldots, n\}$:

- $c'_i \in C'$;
- $i \neq j \rightarrow c'_i \neq c'_j$; and
- $\mathrm{Vars}(\psi_i) \subseteq I \cup \{c'_1, \ldots, c'_{i-1}\} \cup \{c_i, \ldots, c_n\}$. $\quad\square$

Example (Translation into Propositional Logic) The ladder logic formula obtained from figure 5.2 (i) in the naive reading is

$$\psi_n \equiv (e' \leftrightarrow \neg d \vee (c \wedge (a \vee b)))$$

Applying the Tseitin transformation results in

$$
\begin{aligned}
\psi_T \equiv \quad & (x'_5 \leftrightarrow b) \\
\wedge\ & (x'_4 \leftrightarrow a) \\
\wedge\ & (x'_3 \leftrightarrow x'_4 \vee x'_5) \\
\wedge\ & (x'_1 \leftrightarrow c \wedge x'_3) \\
\wedge\ & (x'_2 \leftrightarrow \neg d) \\
\wedge\ & (x'_0 \leftrightarrow x'_1 \vee x'_2) \\
\wedge\ & (e' \leftrightarrow x'_0)
\end{aligned}
$$

Applying the Tseitin transformation gives in larger formulae in this step. $\quad\square$

(Step 2) Specialising General Safety Properties

Transform the given safety property (in a discrete time, temporal first order logic) into a propositional formula using the correct naming schemes for propositional variables, taking into account the given track plan. This results in a safety property φ in propositional logic.

Example (A safety formula) A typical safety property would be: before a route can be set, "*all train detection devices in the route indicate the line is clear*". This is one of the interlocking principles stated in [89]. This principle can be formalized using specialized predicates

- *routeOf* (indicating which routes in a track plan belong to a signal) and
- *tracksOf* (indicating which tracks belong to a route)

which encode the given trackplan, and specialized temporal predicates

- *proceed* (which is true for a signal at a given time, if a train can proceed at this signal),
- *set* (which is true for a route at a given time, if this route is set)

- *occupied* (which is true for a track at a given time, if this track is occupied by a train),

which encode the state associated with the track plan. Using the convention that a primed predicate denotes the next state, the above safety property can be formalized as:

$$\forall s \in Signal, rn \in RouteName, t \in TrackSegment :$$
$$rn \in routesOf(s) \wedge t \in tracksOf(rn) \implies$$
$$((not(proceed(s)) \wedge proceed'(s) \wedge set(rn)) \implies$$
$$(not(occupied(t))))$$

An example collection of predicates that can help to formalize safety properties [85] can be found in section ... on Domain Analysis in this HB-FM-RD.□

(Step 2.1)   Transform the formula into prenex normal form. In this normal form, a formula is written as a string of quantifiers and bound variables, called the prefix, followed by a quantifier-free part, called the matrix. See section ... in this HB-FM-RD for an algorithm. This transformation helps to obtain small safety formulae: after substituting values for the bound variables in (Step 2.2), for verification we can often consider each instantiation of the matrix as a separate formula. This is of advantage as verification time grows with formula size.
Our example above is already in prenex normal form. Safety properties often arise in prenex normal form starting with a universal quantification; thus this step can usually be left out.

(Step 2.2)   Replace all universal and existential quantifiers by appropriate conjunctions and disjunctions, respectively, by using the topological information given through the trackplan. The resulting formula will be variable free, as all variables have been replaced by constant symbols corresponding to the finitely many elements of the track plan.

Example (Illustration of Step 2.2) For the trackplan and the route table in figures 5.3–5.4 and the safety formula in Example 3 we obtain:

$$S100(AM) \in routesOf(S100) \wedge AA \in tracksOf(S100(AM)) \implies$$
$$(not(proceed(S100)) \wedge proceed'(S100) \wedge set(S100(AM))) \implies$$
$$(not(occupied(AA)))$$
$$\wedge\, S100(AM) \in routesOf(S106) \wedge AA \in tracksOf(S100(AM)) \implies$$
$$(not(proceed(S106)) \wedge proceed'(S106) \wedge set(S100(AM))) \implies$$
$$(not(occupied(AA)))$$
$$\wedge\, S100(AM) \in routesOf(S110) \wedge AA \in tracksOf(S100(AM)) \implies$$
$$(not(proceed(S110)) \wedge proceed'(S110) \wedge set(S100(AM))) \implies$$
$$(not(occupied(AA)))$$
$$\wedge\, \ldots$$

Here, *S100(AM)* is a route name, *S100* is a signal name, and *AA* is a track name from the track plan and its route table as shown in figures 5.3–5.4. □

(Step 2.3) Eliminate true premisses; eliminate subformulæ with false premisses. After Step 2.2., the formula consists of a number of subformulae joined by conjunctions. Each of these subformulæ involves an implication using elements of the fixed track plan and its associated route table, relative to which the premise of each and every subformula can be evaluated.

Example (Illustration of Step 2.3) The first subformula in the example illustrating Step 2.2 is:

$$S100(AM) \in routesOf(S100) \land AA \in tracksOf(S100(AM)) \implies$$
$$(not(proceed(S100)) \land proceed'(S100) \land set(S100(AM))) \implies$$
$$(not(occupied(AA)))$$

According to the trackplan and its associated route table, as shown in figures 5.3–5.4, the premiss of this subformula is true (*S100(AM)* is a route that starts at signal *S100* as we can see from the route table; track *AA* belongs to route *S100(AM)* as route *S100(AM)* starts at signal *S100* and ends at signal signal *S104*, c.f. the route table, and track *AA* is on the path from *S100* to *S104* as can be seen on the trackplan). Thus, in Step 2.3 we keep

$$(not(proceed(S100)) \land proceed'(S100) \land set(S100(AM))) \implies$$
$$(not(occupied(AA)))$$

from the first subformula.

By examining another subformula in the same example, a case in which the premiss evaluates to false can be found:

$$S100(AM) \in routesOf(S106) \land AA \in tracksOf(S100(AM)) \implies$$
$$(not(proceed(S106)) \land proceed'(S106) \land set(S100(AM))) \implies$$
$$(not(occupied(AA)))$$

Since route *100(AM)* is not contained within the routes of signal *S106*, the premiss of this subformula is false. Thus, in Step 2.3 we delete the whole subformula. □

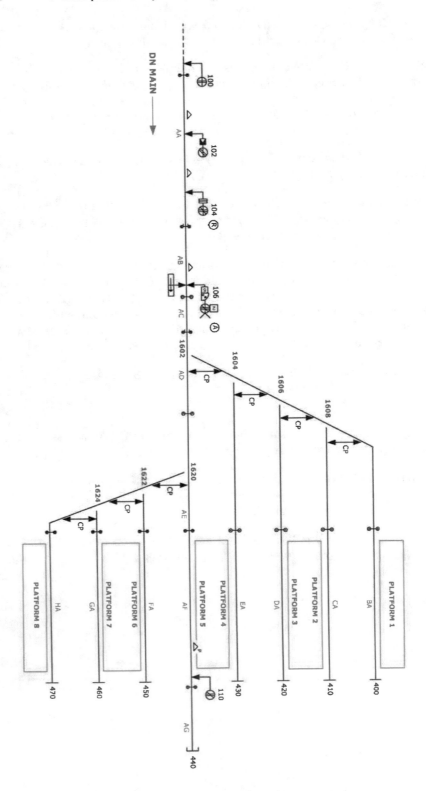

Fig. 5.3 *A track plan, cf. [85].*

Signal Number	Route Letter & Class	Destination		Aspect Type	RI Type JI/SI/MI	RI Legend	Special Notes
		Line	Signal				
S100	A(M)	DN MAIN	S104	M	–	–	–
S106	A(M)	PLATFORM 1	B/STOP 400	M	JI	POS 1	MAR
	A(C)	PLATFORM 1	B/STOP 400	PL	MI	1	–
	B(S)	PLATFORM 2	B/STOP 410	PL	MI	2	NP, P
	C(M)	PLATFORM 3	B/STOP 420	M	SI	3	MAR
	D(S)	PLATFORM 4	B/STOP 430	PL	MI	4	NP
	E(M)	DN MAIN	S110	M	–	–	–
	F(S)	PLATFORM 6	B/STOP450	PL	MI	6	P
	G(M)	PLATFORM 7	B/STOP460	M	SI	7	MAR
	G(C)	PLATFORM 7	B/STOP460	PL	MI	7	–
	H(M)	PLATFORM 8	B/STOP470	M	JI	POS4	MAR
	H(C)	PLATFORM 8	B/STOP470	PL	MJ	8	–
S110	A(M)	DN MAIN	B/STOP440	M	–	–	–

Fig. 5.4 *Part of the Route table associated with the trackplan shown in figure 5.3. A row in the route table describes a single route: in the first column, which signal the route begins at, in the second column what the route name is, in the fourth column which signal the route ends at.*

Element	Property	Prefix	Suffix
Track	Occupied	T<SEGMENT>	.OCC(IL)
Route	Set (Route)	S<SIGNAL(ROUTE)>	.U
Signal	Shows Proceed	S<SIGNAL>	.G

Fig. 5.5 *A typical variable naming scheme, cf. [85].*

(Step 2.4) Replace all predicates with propositional variables according to a variable naming scheme for ladder logic programs.

Example (Illustration of Step 2.4) In the example illustrating Step 2.3 we saw that the subformula

$$(not(proceed(S100)) \wedge proceed'(S100) \wedge set(S100(AM))) \implies (not(occupied(AA)))$$

remained after Step 2.3. Now we replace the state describing predicates with propositional variables:

$$(not(S100.G) \wedge S100.G' \wedge S100(AM).U) \implies (not(TAA.OCC(IL)))$$

To this end, we apply a variable naming scheme such as the one shown in figure 5.5. □

(Step 3): Model Checking

Use model checking to verify if a ladder logic program fulfills (satisfies) a safety property.

(Step 3: Variant *Inductive Model Checking*)

Criteria

Inductive Model Checking is applicable to the subclass *Expected-Success* identified above, i.e., the user expects verification to succeed. Note that for debugging the program, i.e., when one expects the verification to fail, the Variant 'Bounded Model Checking' is better suited as it provides counterexample traces.

Principle

Inductive Model Checking checks if an over-approximation of the reachable state space is safe; for a detailed discussion see the evaluation of the method below.

The set of initial states of the interlocking is often given in the rail industry by setting all output variables to 0 and then to compute one step with Ψ. This can be expressed by a formula *Init*. Furthermore, we need two versions of the safety property φ under discussion: φ itself and a formula φ', which we obtain from φ by adding a prime to all variables.

Inductive Model Checking is then defined by:

---

if $\neg(Init \to \varphi')$ is satisfiable **then**
    return error state
else if $\neg(\psi \wedge \varphi \to \varphi')$ is satisfiable **then**
    return pair of error states
else
    return "safe"

The check if a formula is satisfiable can be done by a SAT solver. Many of those are available as 'off the shelf' software packages; see section ... of this HB-FM-RD for a discussion of SAT solving.

(Step 3: Variant *Bounded Model Checking*)

$$[\dots]$$

(Step 3: Variant *Temporal Induction*)

$$[\ldots]$$

(Step 3: Variant *Stålmarck's Algorithm*)

$$[\ldots]$$

(Step 4): Validating Model Checking Results

(Step 4: Validation for *Inductive Model Checking*)

If verification is successful, the safety property in question has been established and no further validation step is required. However, if verification fails, this result requires interpretation. Failing verification can occur, e.g., due to:

1. incorrect encoding of the safety property in FOL,
2. incorrect use of names of propositional variables,
3. a deliberate deviation from the property in the Ladder Logic program,
4. a false positive,
5. a mistake in the Ladder Logic program.

Note that only two of the listed reasons concern the program, i.e., the system under scrutiny, itself, while three of them address mistakes within other artefacts.

As safety properties and naming conventions remain stable over longer periods of time, these kind of mistakes can be 'eliminated by use', i.e., after 'many' verification attempts have been carried out and the process has been updated accordingly, the proportion of these mistakes will decrease.

A similar argument will apply to the deliberate deviation from the property in the Ladder Logic program: this will not happen only once, but will happen only as an established programming practice. In this case, it would be adequate to change the safety property accordingly, in order to verify that the deviating behaviour has been encoded correctly.

A 'false positive' will arise when the safety property to be verified is not 'inductive', i.e., all reachable states are safe, however, there exists a safe, unreachable state with a transition into an unsafe state (figure 5.6): within the dashed circle we find all possible states of the system, where the green region comprises of all states for which the safety property φ holds; in the beginning, the system is in the state marked by an incoming arrow with no source; the system evolves along the arrows. A 'false positive' is illustrated by the two red states connected with a transition, where one state is safe, the other is not, and the safe state is not reachable from the initial state. Here, Inductive Model Checking will return a false positive, i.e., it will say

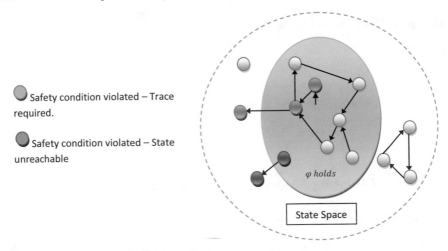

Safety condition violated – Trace required.

Safety condition violated – State unreachable

Fig. 5.6 *State Space Illustration according to [46].*

that the system is unsafe although the safety violation will never be reached in a system run.

One technique to exclude false positives is to add a suitable invariant to the verification. The effect of such an invariant is that it reduces the state space to be considered, hopefully excluding all safe, unreachable states with a transition into an unsafe states. Finding such invariants is a challenge which has been addressed in the literature [19]. For a discussion see section ... of this HB-FM-RD.

Only in the last case there is a need to actually change the Ladder Logic program. It takes experience and (manual) work, to isolate this case from the others.

(Step 4: Validation for *Bounded Model-Checking*)

[...]

(Step 4: Validation for *Temporal Induction*)

[...]

(Step 4: Validation for *Stålmarck's Algorithm*)

[...]

5.3 Academic Explanation

A HB might also include some rationalisation of the method, but not like the one we present in this section. Our rationalisation is meant for an academic audience.

5.3.1 A Short Bibliography

The method of using software model checking of Ladder Logic to verify interlockings is a well established one and belongs to settled knowledge within the railway domain. As early as 1995, Groote et al. utilize it to verify the station Hoorn-Kersenbooger. Their conjecture is that it would be rather straightforward to verify correctness criteria on larger railway yards [32]. Three years later, Fokkink and Hollingshead suggest a systematic translation of Ladder Logic into Boolean formulæ [28]. They conclude: *"The intensive testing procedure for interlockings is time and money consuming, and although the procedure is thorough and carried out by experts and semi-automated simulation, Verification of Interlockings it does not give a 100% guarantee that an interlocking satisfies the dependencies in its control tables"*. Alternative approaches include the work by Zoubek (et al.), who provide a translation from ladder logic into timed automata [91]; this then allows utilisation of the Uppaal model checker as a verification tool.[1] The paper [47], on which we base our presentation here, summarises several industrially funded research projects [51] [52] [46] [64] [85], which lead to several further publications [50] [53] [54].

5.3.2 Solid State Interlockings

In railway systems, solid state interlockings provide a safety layer between the controller and the track. In order to move a train, the (human) controller issues a request to set a route. The interlocking uses rules and track information to determine whether it is safe to permit this request: if so, the interlocking will change the state of the track (move points, set signals, etc.) and inform the controller that the request was granted; otherwise the interlocking will not change the track state. In this sense, an interlocking is like a Programmable Logic Controller (PLC). The standard IEC 61131 [41] identifies programming languages for such controllers, including the visual language Ladder Logic.

---

[1] https://www.uppaal.com/

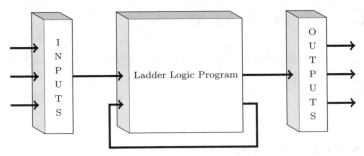

Fig. 5.7 *Control Cycle of a Westrace Interlocking according to [46].*

Closely following James [46], we exemplify the typical interlocking control process by giving details of the three main stages of operation of an interlocking of type 'Westrace'.

Reading of Inputs: The first stage involves reading input values from various sources. The input reading process is undertaken by a specialised 'I/O module'. Inputs may include requests from the controller and details from physical track sensors. It is also possible for input values to be defined as remembered values from the previous execution of the ladder logic program.

Internal Processing: The second stage involves computing new values for output variables. This task uses the variables that have been read in stage one. These variables are then processed by a ladder logic program.

Committing of Outputs: Finally, all calculated outputs are passed back to the 'I/O Module' to be committed to various devices. Here we note that some outputs may actually be remembered by the interlocking, ready to be used within the next execution of the control cycle. In this stage, commands to change the physical track layout may be issued and information may be passed back to the controller.

Interlocking applications such as the Westrace interlocking discussed above are developed according to processes prescribed by railway authorities, in the UK, e.g., by Network Rail's *Governance for Railway Investment Projects* (GRIP) process. The first four GRIP phases define the track plan and routes of the railway to be constructed, while phase five —the detailed design— is contracted to a signalling company which chooses appropriate track equipment, adds control tables to the track plan, and implements the solid state interlocking. It is for part of this phase, namely for the correct implementation of a control table in a solid state interlocking, that we described support in terms of a Formal Method.

The problem addressed is the verification of safety properties for railway interlocking programs written in Ladder Logic, a process prescribed in standards and by authorities. Figure 5.8 shows the standard approach as usually taken in industry as well as the suggested, improved process. In the original

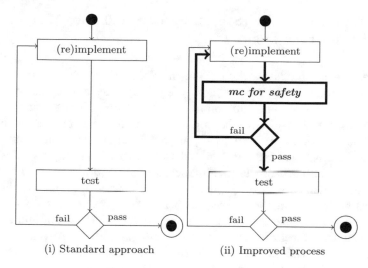

Fig. 5.8 *Process Improvement by including model checking (mc) for Safety.*

process (i), the control program is implemented in Ladder Logic and tested afterwards. In case the program fails a test, re-implementation is necessary. In industrial practice, such an implement-test cycle is iterated about five to ten times until the interlocking program finally passes all tests. The improved process (ii) involves a lightweight application of a Formal Method: testing takes place only if model checking for safety succeeds.

5.3.3 Explanation for (Step 1): Tseitin Transformation

Using the Tseitin transformation [82] to translate a Ladder Logic program into propositional logic introduces a proxy variable for each subformula. This leads to a formula which is equisatisfiable to the naive reading, i.e., the former is satisfiable exactly if the latter is satisfiable. However, Tseitin's formula is not equivalid as it has new variables.

The advantage of the Tseitin transformation becomes apparent when one transforms the result into CNF. The naive approach reads the Boolean expression of a rung and uses De Morgan's laws and the distributive property to convert it to CNF. This can result in an exponential increase in size. In contrast, the Tseitin transformation outputs formulae whose size grows only linearly relative to the size of the original expression.

5.3.4 Explanation for (Step 2): Discrete Time, Temporal First Order Logics

The presentation of the method given in **(Step 2)** for this step closely follows [85]. In order to capture generic safety properties and to adapt them to specific trackplans, we are using many sorted first order logics over states and their successors. These logics comprise of elements that enable the expression of the generic safety properties, as well as of elements that correspond to a specific trackplan with associated route table —see figures 5.3–5.4 for an example— and a variable naming scheme; see figure 5.5 for an example. Having both elements in one logic, the generic ones and the specific ones, allows us to transform the generic properties into specific ones. As also the models are formed specific to the trackplan, route table, and variable naming scheme, the transformation steps given in **(Step 2)** above are semantics preserving, i.e., the original formula and the transformed formula are equivalent in the logics. Below, we provide a sketch of these logics.

Signatures

Signatures are many sorted, first-order logic signatures (S, F, P) where

- the set of sort symbols $S = \{ TrackSegment, Signal, RouteName, Point \}$;
- the set F of function symbols consists of

$$\{ \; routesOf : Signal \to \mathcal{P}(RouteName),$$
$$pointsOf : RouteName \to \mathcal{P}(SetofPoint)$$
$$tracksOf : RouteName \to \mathcal{P}(TrackSegment)\},$$

  as well as a number of constants arising from the trackplan, such as $S100 : Signal$;
- and where the set P of predicate symbols consists of

  - a number of unary and binary predicates symbols that relate to the state to the trackplan, such as $isInCorrectPositionFor : Point \times RouteName$ and $proceed : Signal$ and
  - and a number of propositional variables generated for the given track plan according to the variable naming scheme, e.g., $S100.G$.

  See [85] for the complete list of predicate symbols. Some of the unary predicate symbols and the propositional variables are connected through equivalences, e.g., $proceed(S100) \iff S100.G$.

In such a signature, sorts correspond to the families of elements a trackplan refers to, functions enable us to encode topological information, and predicates encode the state of the stateful elements of a trackplan.

Models

Models at a point of time are pairs (T, I) where

- T is a track plan
- I is a propositional model for all propositional variables associated with T, e.g., $I(S106.G) \in \{true, false\}$.

Sorts and functions, are given a fixed interpretation according to the track plan, e.g.,

- $Signal_T = \{S100, ...\}$. iff T *has signals* $S100, ...$
- $routesOf_T(s) = \{r_1, ..., r_n\}$: iff *in* T, *signal* s *has routes* $r_1, ..., r_n$

Predicates obtain their interpretations from a combination of looking up information from both the trackplan T and the propositional model I:

- p $isInCorrectPositionFor_{T,I}$ r *holds* iff

  - **case 1:** in T, p needs to be in reverse for r and $I(p.RL)$ is true
  - **case 2:** in T, p needs to be in normal for r and $I(p.NL)$ is true

The models of a signature are then sequences of the form $(T, I_0), (T, I_1), ...,$ i.e., the trackplan in the first component stays constant, only the state of the propositional variables is changing.

Formulæ

The formulae are standard first order logic formulæ, where predicate symbols can also appear with a prime, e.g., $S106.G$ and $S106.G'$ are both atomic formulæ. The prime indicates that a predicate shall be evaluated in the successor state.

Satisfaction

Given two models $(T, I_1), (T, I_2)$, satisfaction of a formula is satisfaction as in first order logic, where unprimed predicates are evaluated over (T, I_1) and primed predicates are evaluated over (T, I_2). A formula φ holds in a sequence $\langle (T, I_0), (T, I_1), ... \rangle$, iff for all $i \geq 0$ the formula φ holds over $(T, I_i), (T, I_{i+1})$.

5.3.5 Explanation of Step (3): Verification Problem associated with Ladder Logic

One can associate an automaton with a ladder logic formula. Recall that the representation of a ladder logic program in propositional logic is constructed in terms of disjoint finite sets I and C of input and output variables, where internal variables are subsumed in C. This automaton has interpretations of the set of propositional variables $I \cup C$ as its states, i.e., the configurations of the PLC. In order to define the automaton's transition relation, we introduce paired valuations. Here, the function *unprime* deletes the prime from a variable.

Definition (Paired Valuations) Given a finite set of input variables I, a finite set of state variables C, and valuations $\mu, \nu : (I \cup C) \to \{0, 1\}$ we define the paired valuation $\mu \, \S \, \nu : (I \cup C \cup I' \cup C') \to \{0, 1\}$ where

$$\mu \, \S \, \nu(x) = \begin{cases} \mu(x) & \text{if } x \in I \cup C \\ \nu(unprime(x)) & \text{if } x \in I' \cup C'. \end{cases} \qquad \square$$

The automaton corresponding to a ladder logic formula is defined as follows:

Definition (Automaton) Given a ladder logic formula ψ over $I \cup C$, we define the automaton

$$A(\psi) = (S, S_0, \to)$$

where

- $S = \{\mu \mid \mu : I \cup C \to \{0, 1\}\}$ is the set of states,
- $\mu \xrightarrow{\nu(I')} \nu$ if $\mu \, \S \, \nu \models \psi$, defines the transition relation, and
- $S_0 = \{\nu \mid \exists \mu : \mu \models \neg C, \mu \, \S \, \nu \models \psi\}$ gives the set of initial states. Here, $\neg C$ expands to $\bigwedge_{c \in C} \neg c$. $\qquad \square$

Having associated an automaton with a ladder logic program, one can now define the verification problem:

Definition (Verification Problem for Ladder Logic Programs) $A(\psi) \models \varphi$ iff for all reachable states μ of $A(\psi)$ we have $\mu \models \varphi$. $\qquad \square$

This verification problem can be addressed through various model checking approaches.

5.4 Experience Reports Concerning Step 6

The paper [47] was written with a view to sum up the Swansea Railway Verification Group's knowledge on the topic in HB style — however, without having the above presentation scheme at hand.

From a writing point of view, the experience was that the scheme offers a clear guideline of what to describe and at which point. This led to many differences in the presentation:

- Most prominent is the shift of perspective towards engineering: while our paper [47] is driven by scientific questions, the main focus of the presentation in the sections 5.1–5.2 is on which steps the practitioner would have to undertake in order to build an automated system.
- Thanks to the scheme's concise questions, discussions of certain topics were added: the 'subclasses' discussed were not a topic in [47], the explicit need to address 'gains' led to a stronger discussion of the point, the same holds for the 'validation' section.

Though the scheme's questions required to put more effort in the discussion of topics that were neglected in [47], overall the writing was easier and, in comparison, took less time.

Subsequently, test readers unanimously preferred the above handbook entry over our paper [47] and said that —thanks to the scheme— the discussion of the topic has become much clearer: the reader knows what to expect where; overall the argument is better structured. The 'step by step' approach was applauded and readers felt 'better informed'.

Though the scale of our experiment is small, it indicates that the suggested presentation scheme is a step in the right direction.

Chapter 6
Conclusions and Prospects for Future Work

A *practitioner's handbook* (HB) about the *application of Formal Methods* of computing to a particular domain —in our case: the railway domain [9]— must *bridge the gap* between Formal Methods (which are closely related to theoretical computer science), as such, and their practical applicability by those engineers who have perhaps never studied theoretical computer science in all its scholarly details. This 'bridge' between Formal Methods, as such, and their practical application was called 'formal *engineering* methods' (FEM) by Shaoying Liu [65](p. 11: figure 1.5), who has rightly emphasised the need for helpful additional *"methods that support the application of Formal Methods to the development of large-scale computer systems"* [65](p. 10). Thus, FEM *"are equivalent neither to application of Formal Methods, nor to Formal Methods themselves. They are intended to serve as a bridge between Formal Methods and their applications, providing methods and related techniques to incorporate Formal Methods into the entire software engineering process. Without such a bridge, application of Formal Methods is difficult. The quality of the bridge may affect the smoothness of Formal Methods technology transfer. Some types of bridges may make the transfer easier than others, so the important point is how to build the bridge"* [65](pp. 10-11).

In the 'spirit' of Liu's above-quoted words about FEM, we have motivated and described in this book a systematic meta method for HB construction, with particular regard to the field of Formal Methods in the railway domain. For this purpose, the following elements are crucial:

- Clarification of what shall be understood by the term 'HB',
- Clarification of what shall be regarded as 'settled knowledge',
- Choice of a suitable historic data base in which we can reasonably expect to find settled knowledge,
- Extraction and interpretation of the relevant information from the chosen data base by means of Formal Concept Analysis (FCA), and
- Syntactic-structural transformation (a.k.a. 'HB-ification') of the selected contents into an HB compatible form of presentation that is problem so-

© The Author(s), under exclusive license to Springer Nature Switzerland AG 2020
S. Gruner et al., *On the Construction of Engineering Handbooks*, SpringerBriefs
in Computer Science, https://doi.org/10.1007/978-3-030-44648-2_6

lution oriented (in a 'cook book' style with 'recipes') for the benefit of industrial practitioners.

We have devised a six step HB construction method, which we have *illustrated* by means of a reasonably large *example* (case study). We do *not* claim that our illustrative example is already 'complete' or 'comprehensive enough' for the immediate composition of an envisaged FM-Railway HB. At this point we are also able to see some 'overlaps' between our work and the broader discipline of 'information science'.[1]

Based on the methods and results described and discussed in this book, an array of future tasks supporting the publication of a useful HB about the application of Formal Methods in the railway domain still remains to be tackled:

- Continuation and 'deepening' of our information scientific studies of 'settled' knowledge (particularly of Formal Methods, particularly in the railway domain) where our reported example is not yet comprehensive enough;
- Definite decision concerning the knowledge topics to be included into the future HB on the chosen theme;
- Systematic (and more comprehensive) application (including several finer details) of our HB-ification meta method to the selected knowledge 'items';
- Publication and release (with help of many additional experts) of the envisaged HB on the application of Formal Methods in the railway domain;
- Appropriate conceptual and methodical 'placement' of the HB into its appropriate position 'between' the general (high-level) industrial norms and standards (such as, for example, CENELEC or ISO-26262) on the one hand, and the business specific individual micro practices of every individual engineer in the industry, on the other hand, in such a manner that the HB can rightfully be regarded as a helpful (neither too abstract nor too case specific) 'compliance aid' w.r.t. to those over arching industrial standards;
- Further iterations of the same research cycle for the sake of the still absent HB-ifications of several other fields in which Formal Methods are also applied, such as: automobile engineering, avionics, etc.

More generally, the problem of HB construction in engineering needs to be considered. This is important as new Engineering subdisciplines arise at an accelerating rate, for example, the domain of cyber physical systems. As these new domains are fast growing, waiting for the traditional evolutionary approach to HB construction will prove to be unsatisfactory. Here, out meta method may be of some help.

---

[1] https://en.wikipedia.org/wiki/Information_science

References

1. C. Alexander, S. Ishikawa, M. Silverstein, M. Jacobson, I. Fiksdahl-King, S. Angel, *A Pattern Language: Towns, Buildings, Construction.* Oxford University Press, 1977.
2. B. Alpern, F. Schneider, *Defining Liveness.* Information Processing Letters 21/4, pp. 181-185, 1985.
3. A. Arageorgis, A. Baltas, *Demarcating Technology from Science: Problems and Problem Solving in Technology.* Zeitschrift für allgemeine Wissenschaftstheorie 20/2, pp. 212-229, 1989.
4. M. ter Beek, A. Borälv, A. Fantechi, A. Ferrari, S. Gnesi, C. Löfving, F. Mazzanti, *Adopting Formal Methods in an Industrial Setting: The Railways Case.* LNCS 11800, pp. 762-772, 2019.
5. M. ter Beek, S. Gnesi, A. Knapp, *Formal Methods for Transport Systems.* International Journal on Software Tools for Technology Transfer 20/3, pp. 237-241, 2018.
6. U. Berger, P. James, A. Lawrence, M. Roggenbach, M. Seisenberger, *Verification of the European Rail Traffic Management System in Real-Time Maude.* Science of Computer Programming 154, pp. 61-88, 2018.
7. M. Bhatti, N. Anquetil, M. Huchard, S. Ducasse, *A Catalog of Patterns for Concept Lattice Interpretation in Software Reengineering*, pp. 118-123 in: Proceedings 24th International Conference on Software Engineering and Knowledge Engineering, 2012.
8. G. Birkhoff, *Lattice Theory.* 2nd ed., American Math. Soc., 1948.
9. D. Bjørner, *Formal Software Techniques in Railway Systems.* IFAC Proceedings Volumes 33/9, pp. 101-108, 2000.
10. D. Bjørner, *Software Engineering, Vol. 1: Abstraction and Modelling.* Springer-Verlag, 2006.
11. D. Bjørner, K. Havelund, *40 Years of Formal Methods: Some Obstacles and Some Possibilities?* LNCS 8442, pp. 42-61, 2014.
12. J. Bowen, M. Hinchey, *Formal Methods*, ch. 71 in Computing Handbook, Vol. 1: Computer Science and Software Engineering, 3rd ed. CRC Press, 2014.
13. J. Bowen, S. Reeves, *From a Community of Practice to a Body of Knowledge: A Case Study of the Formal Methods Community.* LNCS 6664, pp. 308-322, 2011.
14. M. Bunge, *Philosophy of Science, Vol. 2: From Explanation to Justification.* Revised ed., Transaction Publ., 1998.
15. P. Burmeister, *ConImp — Ein Programm zur Formalen Begriffsanalyse*, pp. 25-56 in: Begriffliche Wissensverarbeitung: Methoden und Anwendungen. Springer-Verlag, 2000.
16. P. Burmeister *Formal Concept Analysis with ConImp: Introduction to the Basic Features*, (English transl. of [15]). Technical Report: Fachbereich Mathematik, Technische Universität Darmstadt, 2003.

17. A. Buzmakov, S. Kuznetsov, A. Napoli, *Is Concept Stability a Measure for Pattern Selection?* Procedia Computer Science 31, pp. 918-927, 2014.

18. K. Chan, *Formal Methods for Web Services: a Taxonomic Approach*, pp. 357-360 in: 32nd International Conference on Software Engineering (Vol. 2). IEEE, 2010.

19. A. Cimatti, A. Griggio, S. Mover, S. Tonetta, *Infinite-State Invariant Checking with IC3 and Predicate Abstraction.* Formal Methods in System Design 49/3, pp. 190–218, 2016.

20. E. Clarke, J. Wing, *Formal Methods: State of the Art and Future Directions.* ACM Computing Surveys 28/4, pp. 626-643, 1996.

21. E. Constant, *The Origins of the Turbojet Revolution.* Johns Hopkins Studies in the History of Technology 5, Johns Hopkins University Press, 1980.

22. A. Cooper, D. Kourie, S. Coetzee, *Thoughts on Exploiting Instability in Lattices for Assessing the Discrimination Adequacy of a Taxonomy*, art. 4 in: Proceedings CLA'10 Concept Lattices and their Applications, 2010.

23. E. Currás, *Ontologies, Taxonomies and Thesauri in Systems Science and Systematics.* Chandos Publ., 2010.

24. P. Duggan, A. Borälv, *Mathematical Proof in an Automated Environment for Railway Interlockings.* IRSE News 217, pp. 2-6, 2015.

25. B. Everitt, S. Landau, M. Leese, *Cluster Analysis.* 4th ed., Arnold publ., 2001.

26. A. Fantechi, *Twenty-Five Years of Formal Methods and Railways: What Next?* LNCS 8368, pp. 167-183, 2014.

27. L. Fleck, *Entstehung und Entwicklung einer wissenschaftlichen Tatsache.* Benno Schwabe & Co. Publ., 1935. English transl.: *Genesis and Development of a Scientific Fact.* University of Chicago Press, 1979.

28. W. Fokkink, *Verification of Interlockings: From Control Tables to Ladder Logic Diagrams.* Invited lecture: Third International Workshop on Formal Methods for Industrial Critical Systems, 1998.[2]

29. B. Ganter, R. Wille, *Formale Begriffsanalyse: Mathematische Grundlagen.* Springer-Verlag, 1996.

30. H. Garavel, S. Graf, *Formal Methods for Safe and Secure Computer Systems.* Technical Report: BSI Study 875, Bundesamt für Sicherheit in der Informationstechnik, Federal Republic of Germany, 2013.

31. M. Gleirscher, D. Marmsoler, *Formal Methods: Oversold? Underused? A Survey.* Technical Report: arXiv:1812.08815 [cs.SE], 2018.

32. J. Groote, S. van Vlijmen, J. Koorn, *The Safety-Guaranteeing System at Station Hoorn-Kersenboogerd*, pp. 57-68 in: COMPASS'95 Proceedings of 10th Annual Conference on Computer Assurance Systems Integrity, Software Safety and Process Security. IEEE, 1995.

33. S. Gruner, *On the Historical Semantics of the Notion of 'Software Architecture'.* Journal for Transdisciplinary Research in Southern Africa 10/1, pp. 37-66, 2014.

34. S. Gruner, A. Haxthausen, T. Maibaum, M. Roggenbach, *FM-RAIL-BOK Organizers' Message.* LNCS 8368, pp. XI-XII, 2014.

35. S. Gruner, A. Haxthausen, T. Maibaum, M. Roggenbach, *Towards a Formal Methods Body of Knowledge for Railway Control and Safety Systems: FM-RAIL-BOK Workshop 2013.* Technical Report: Technical University of Denmark, 2013.[3]

36. S. Gruner, A. Kumar, T. Maibaum, *Towards a Body of Knowledge in Formal Methods for the Railway Domain: Identification of Settled Knowledge.* CCIS 596, pp. 87-102, 2016.

37. A. Haxthausen, H. Nguyen, M. Roggenbach *Comparing Formal Verification Approaches of Interlocking Systems*, pp. 160-177 in: Proceedings 1st International Conference on Reliability, Safety, and Security of Railway Systems, 2016.

---

[2] http://fmics.inria.fr/workshop-3/

[3] https://ssfmgroup.wordpress.com/rel/

38. A. Haxthausen, J. Peleska, *Formal Development and Verification of a Distributed Railway Control System.* LNCS 1709, pp. 1546-1563, 1999; and also IEEE Transactions on Software Engineering 26/8, pp. 687-701, 2000.
39. M. Heidegger, *Sein und Zeit.* Niemeyer-Verlag, 1927.
40. M. Hinchey, M. Jackson, P. Cousot, B. Cook. J. Bowen, T. Margaria, *Software Engineering and Formal Methods.* Communications of the ACM 51/9, pp. 54-59, 2008.
41. IEC, *IEC 61131-3 edition 2.0 2003-01.* International Standard: Programmable Controllers, Part 3: Programming Languages. 2003.
42. M. Ikeda, A. Yamamoto, *Classification by Selecting Plausible Formal Concepts in a Concept Lattice,* pp. 22-35 in: Proceedings FCAIR2013 Workshop on Formal Concept Analysis meets Information Retrieval, 2013.
43. M. Jackson, *Formal Methods and Traditional Engineering.* Journal of Systems and Software 40/3, pp. 191-194, 1998.
44. M. Jackson, *The Name and Nature of Software Engineering.* LNCS 5316, pp. 1-38, 2008.
45. M. Jackson, *The Operational Principle and Problem Frames,* pp. 143-165 in: Reflections on the Work of C.A.R. Hoare. Springer-Verlag, 2010.
46. P. James, *SAT-based Model Checking and its Applications to Train Control Software.* Master of Research thesis: Swansea University, 2010.
47. P. James, A. Lawrence, F. Moller, M. Roggenbach, M. Seisenberger, A. Setzer, K. Kanso, S. Chadwick, *Verification of Solid State Interlocking Programs.* LNCS 8368, pp. 253-268, 2014.
48. P. James, F. Moller, H. Nguyen, M. Roggenbach, S. Schneider, H. Treharne, *On Modelling and Verifying Railway Interlockings: Tracking Train Lengths.* Science of Computer Programming 96, pp. 315-336, 2014.
49. P. James, F. Moller, H. Nguyen, M. Roggenbach, S. Schneider, H. Treharne, *Techniques for Modelling and Verifying Railway Interlockings.* International Journal on Software Tools for Technology Transfer 16/6, pp. 685-711, 2014.
50. P. James, M. Roggenbach, *Automatically Verifying Railway Interlockings using SAT-based Model Checking.* ECEASST 35, art. 10, 2010.
51. K. Kanso, *AGDA as a Platform for the Development of Verified Railway Interlocking Systems.* Doctoral dissertation, Swansea University, 2013.
52. K. Kanso, *Formal Verification of Ladder Logic.* Master of Research dissertation: Swansea University, 2010.
53. K. Kanso, F. Moller, A. Setzer, *Automated Verification of Signalling Principles in Railway Interlocking Systems.* Electronic Notes in Theoretical Compututer Science 250/2, pp. 19-31, 2009.
54. K. Kanso, A. Setzer, *A light-weight Integration of Automated and Interactive Theorem Proving.* Mathematical Structures in Computer Science 26/1, pp. 129-153, 2016.
55. M. Kezadri, M. Pantel, *First Steps toward a Verification and Validation Ontology,* pp. 440-444 in: Proceedings of the International Conference on Knowledge Engineering and Ontology Development, 2010.
56. D. Knuth, *The Art of Computer Programming, Vol. 3: Sorting and Searching.* 2nd ed., Addison-Wesley, 1998.
57. T. Kolesnykova, O. Matveyeva, L. Manashkin, M. Mishchenko, *Railway Transportation of Dangerous Goods: a Bibliometric Aspect,* art. 03014 in: 2nd International Scientific and Practical Conference on Energy-Optimal Technologies, Logistic and Safety on Transport. MATEC Web of Conferences 294, 2019.
58. T. Kuhn, *The Structure of Scientific Revolutions.* University of Chicago Press, 1962.
59. A. Kumar, *A Preparatory Study Towards a Body of Knowledge in the Field of Formal Methods for the Railway Domain.* Master of Applied Science thesis: McMaster University, 2015.
60. S. Kuznetsov, *On Stability of a Formal Concept.* Annals of Mathematics and Artificial Intelligence 49/1-4, pp. 101-115, 2007.

61. S. Kuznetsov, D. Ignatov, *Concept Stability for Constructing Taxonomies of Web-site Users.* Technical Report: arXiv:0905.1424 [cs.CY], 2009.

62. S. Kuznetsov, T. Makhalova, *On Interestingness Measures of Formal Concepts.* Information Sciences 442-443, pp. 202-219, 2018.

63. A. van Lamsweerde, *Formal Specification: a Roadmap.* pp. 147-159 in: Proceedings of the Conference on the Future of Software Engineering. ACM, 2000.

64. A. Lawrence, *Verification of Railway Interlockings in SCADE.* Master of Research thesis, Swansea University, 2011.

65. S. Liu, *Formal Engineering for Industrial Software Development.* Springer-Verlag, 2004.

66. B. Luteberget, C. Johansen, *Efficient Verification of Railway Infrastructure Designs against Standard Regulations.* Formal Methods in System Design 52/1, pp. 1-32, 2018.

67. T. Maibaum, *Formal Methods versus Engineering.* ACM SIGCSE Bulletin 41/2, pp. 6-12, 2009.

68. T. Maibaum, *Mathematical Foundations of Software Engineering: a Roadmap,* pp. 161-172 in: Proceedings of the Conference on the Future of Software Engineering. ACM, 2000.

69. T. Maibaum, *What is a BoK? (Extended Abstract).* LNCS 8368, pp. 184-188, 2014.

70. F. Mazzanti, A. Ferrari, G. Spagnolo, *Towards Formal Methods Diversity in Railways: an Experience Report with Seven Frameworks.* International Journal on Software Tools for Technology Transfer 20/3, pp. 263-288, 2018.

71. N. Meddouri, M. Meddouri, *Classification Methods based on Formal Concept Analysis,* pp. 9-16 in: Proceedings 6th International Conference on Concept Lattices and their Applications, 2008.

72. J. Oliveira, *A Survey of Formal Methods Courses in European Higher Education.* LNCS 3294, pp. 235-248, 2004.

73. M. Polanyi, *Personal Knowledge: Towards a Post-Critical Philosophy.* Routledge & Kegan Paul, 1958.

74. H. Poser, *On Structural Differences between Science and Engineering.* Techné: Research in Philosophy and Technology 4/2, pp. 128-135, 1998.

75. O. Prokasheva, A. Onishchenko, S. Gurov, *Classification Methods based on Formal Concept Analysis,* pp. 95-104 in: Proceedings FCAIR2013 Workshop on Formal Concept Analysis meets Information Retrieval, 2013.

76. G. Rogers, *The Nature of Engineering: A Philosophy of Technology.* Macmillan International Higher Education, 1983.

77. C. Roth, S. Obiedkov, D. Kourie, *Towards Concise Representation for Taxonomies of Epistemic Communities.* LNAI 4923, pp. 240-255, 2008.

78. M. Shaw, *The Coming-of-Age of Software Architecture Research,* pp. 656-663 in: Proceedings 23rd ICSE. IEEE, 2001.

79. D. Siefkes, *Sinn im Formalen?,* pp. 97-114 in: Sichtweisen der Informatik. Vieweg-Verlag, 1992.

80. D. Smith, *KIDS — A Knowledge-Based Software Development System,* pp. 483-514 in: Automating Software Design. MIT Press, 1991.

81. K. Taguchi, H. Nishihara, T. Aoki, F. Kumeno, K. Hayamizu, K. Shinozaki, *Building a Body of Knowledge on Model Checking for Software Development,* pp. 784-789 in: Proceedings 37th Annual Computer Software and Applications Conference. IEEE, 2013.

82. G. Tseitin, *On the Complexity of Derivation in Propositional Calculus,* pp. 466-483 in: Automation of Reasoning Vol. 2: Classical Papers on Computational Logic 1967-1970. Springer-Verlag, 1983.

83. S. Vanit-Anunchai, *Modelling and Simulating a Thai Railway Signalling System using Coloured Petri Nets.* International Journal on Software Tools for Technology Transfer 20/3, pp. 243-262, 2018.

84. W. Vincenti, *What Engineers know and How they know it: Analytical Studies from Aeronautical History*. Johns Hopkins Studies in the History of Technology 11, John Hopkins University Press, 1990.

85. T. Werner, *Safety Verification of Ladder Logic Programs for Railway Interlockings*. Master of Research thesis, Swansea University, 2017.

86. R. Wille, *Concept Lattices and Conceptual Knowledge Systems*. Computers & Mathematics with Applications 23/6-9, pp. 493-515, 1992.

87. J. Wing, *A Specifier's Introduction to Formal Methods*. IEEE Computer 23/9, pp. 8-23, 1990.

88. J. Woodcock, P. Gorm-Larsen, J. Bicarregui, J. Fitzgerald, *Formal Methods Practice and Experience*. ACM Computing Surveys 41/4, pp. 1-36, 2009.

89. P. Woolford, *Interlocking Principles, Railway Group Standard GK/RT0060*. Technical Report: Rail Safety and Standards Board, 2003.

90. S. Yevtushenko, *System of Data Analysis — Concept Explorer*, pp. 127-134 in: Proceedings 7th National Conference on Artificial Intelligence (KII'00), 2000.

91. B. Zoubek, J. Roussel, M. Kwiatowska, *Towards Automatic Verification of Ladder Logic Programs*. Proceedings IMACS-IEEE Computational Engineering in Systems Applications (CESA'03), 2003.

Printed in the United States
By Bookmasters